In One Barn

Efficient Livestock Housing and Management

By Lee Pelley

The Countryman Press

Maxham Meadow • Woodstock • Vermont 05091

Illustrations by Jeffrey Scott and Katharine Dodge
Cover, text, and reference symbols designed
by Leslie Stevenson Fry

Library of Congress Cataloging in Publication Data

Pelley, Lee.
In one barn.

Bibliography: p. 147
Includes index.
1. Livestock—Handbooks, manuals, etc. 2. Livestock—Housing—Handbooks, manuals, etc. I. Title.
SF65.2.P45 1984 636.08'31 83-25219
ISBN 0-88150-006-2 (pbk.)

Manufactured in the United States of America

For Jeff whose myriad talents and compassions help many . . . but especially me.

PREFACE

In the course of teaching people at our homesteading school and from visiting other homesteads, I have become acutely aware of the lack of planning in most small-scale, mixed-species barns. To add to the problem, housing inefficiencies may accumulate into monstrous difficulties in a short time. Once problems become established, they are very hard to straighten out.

In the shuffle of distractions, the steady growth in the number of animals often gets overlooked. When problems begin to accumulate over time, the day inevitably comes when you find yourself climbing through the buck pen to get to the hay loft, or by-passing hungry animals in your haste to feed those further down the line. And you wonder how this happened to you.

The same competitive instinct that makes animals gain weight quickly when they are in the company of their own kind, has the opposite effect when animals can see their food, but cannot reach it. Such psychological factors in animal housing are of vital importance, but they usually receive the least attention.

Stress on the humans caring for the animals should not be ignored, either. We could *all* do with a little less stress. Stress doesn't improve the health of any living thing.

Solutions to housing problems aren't always obvious. Occasionally, awareness of the problem is just an inner conviction that "this could be done better" rather than an understanding of what the problem is. And because it is most difficult to see things right under your nose, homesteaders with years of experience frequently have as many problems as those just starting out. If we have only one thing in common, it is that we are all individualists and like to do things our own way. Which is precisely what makes us "different"—special, if you will.

I have written this book *not* to instruct you on how to build your barn or operate the one you have now, but instead to present methods of animal housing you may not have thought of yourself. The book also includes information on the different species in case you're thinking of expanding.

I encourage you to read the entire book before beginning to use it. Much of the information about—say goats—would be helpful to the horse or cattle raiser.

For the reader who is contemplating a barn and has yet to make a commitment, I envy you. Had I made fewer mistakes, this book might never have been written. But, since I did make mistakes, I hope you may avoid these same errors, as well as some of your own invention.

INTRODUCTION

Specialization is for insects.

ROBERT HEINLEIN
in TIME ENOUGH FOR LOVE

SPECIALIZATION VERSUS DIVERSITY

We live in a society where specialization is the norm. Consequently, diversity has become associated with incompetence; it implies you are a dilletante at everything.

Specialization has greatly benefitted our society. It has been responsible for great advances in medicine, housing, food production, and the decrease in many adversities such as poverty and hunger.

For the small-scale farmer the benefits of specialization are substantial. Equipment needs are fewer for a single species. Housing can be simplified to accommodate the needs of only one kind of animal. Extensive knowledge about one species is easier to acquire than about many. Specialization also means marketing your farm products to fewer, if more than one, market source. And learning the necessary market skills comes faster if you have only one or two products to sell.

Diversification means many products to sell, so you must learn to deal with a variety of markets. It means you must acquire more diversified knowledge, and work harder to understand each species and its special needs. More thought must go into housing, including concerns not a problem for the single-species manager, such as diseases which can be transmitted from one species to another. You will need more and varied equipment, and more talent in dealing with the variety of animals on a daily basis. In addition, you'll need to know how species interact under a single roof.

A variety of your own products can also mean less dependence on the marketplace—especially on a single market that, if it dried up, would leave you without a buyer for a period of time. Raising different animal species also allows you and yours to live on a more varied diet, no matter what the market is doing. While diseases that cross from one species to another can become a serious problem, you are far less likely to be completely wiped out than the single-species manager. And in nature, plants and animals intermix and do very well. Basically, diversification is what organic gardening and farming are all about; it's environmentally sound to grow animals on the same principle.

And what about you? You are a major factor in this issue. You're probably interested in small-scale farming because you have many talents or would like to acquire them. You don't want to be a carpenter, a veterinarian, a cook, a food grower; you want to be *all* of these things . . . and more. You want less dependence and a more complete, rounded life. You want to see the big picture—the similarities and differences between yourself and your animals on a larger scale than specialization will allow. You want to eliminate as much waste in your life as you can. You want to be proficient at many things instead of excellent at one and incompetent at everything else. And you know it's worth your while to do these things.

Under One Roof

Combining species in one barn can be a health hazard. Interspecies diseases are only a problem under such conditions. Too many people either ignore or don't even realize they are running a risk. Ignorance is an expensive luxury. I'm not speaking of a necessarily virulent disease that may be fatal. Instead, it may be something like ringworm spreading through a variety of animals; some animals will be easily cured, others not. And those easily cured will be reinfected by those who are still diseased. It can be a round-robin of seemingly unstoppable proportion. Prevention is always the best medicine.

Still, there are many valid reasons for joining a variety of species under a single roof. Chores are more centralized, which reduces your running hither and yon to get everyone fed, watered, bedded, and cleaned. All supplies can be kept in this central location, and no duplication is necessary such as of grooming brushes or medicines. Heat is more easily retained, work is reduced, and the quality of care goes up when all your stock are under a single roof.

Not too many people realize that one barn may also result in lower taxes; taxes are often based on the number of roofs under a given ownership. Three sheds would be counted as roofs with the taxes often amounting to more than those for a single barn.

On Housing Fowl

There are, however, some animals who should *not* be housed under the same roof. As a rule fowl do not belong in mammal housing. Disease potential is much greater when fowl and mammals are under the same roof. Even a pseudo-disease problem such as lice can raise havoc in your mammals.

Tuberculosis, which your dairy animals should be tested for yearly, is a species-specific disease. In other words, it *cannot* cross from one species to another. A positive test to tuberculin does *not* necessarily mean that the animal has the disease; it can mean the animal is a carrier of a type belonging to another species—the positive animal may not even be sick. But it's difficult, if not impossible to isolate the disease to prove it. And avian tuberculosis, carried by poultry, may make all of your animals test positive in shared housing.

It's the law that animals testing positive must be slaughtered to prevent the spread of such an economically and physically dangerous disease. You cannot test for it yourself, either; it must be done by a competent veterinarian.

There are other problems in housing mammals and fowl together. Fowl carry lice and, while some mammalian species carry lice anyway, chicken lice can be particularly difficult to get rid of. If caged rabbits are in the barn, they may be subjected to intense itching that they can't escape. Water birds love to slop water when they drink which makes it darn hard to keep the mammal housing from becoming damp.

All birds flap their wings which startles many mammals. Some poultry climb mangers to roost in higher places, a natural enough instinct. I've seen too many cases where mammals were so alarmed by this they avoided the mangers. Others stay away because the fowl contaminate their foodstuffs. Although all species have some form of aggression, larger animals often lose to smaller ones; to attack, ducks and geese peck and pinch and can get downright abusive of larger animals. In small quarters, they have the ability to move around more freely than the larger animals. Can you imagine a horse successfully kicking a duck? Or a goat butting an abusive goose? It's often the natural order for the smaller and more agile to win.

I love geese, but I have seen them successfully corner groups of larger animals. Once they've done it, it's impossible to convince them they shouldn't. Although the mammals may figure out ways around them, the geese just don't give up. I once saw a gander take on a horse. The disagreement ended quickly—wings one, hooves nothing.

If water birds find out mammals have their own water supply, they'll spend all day, if necessary, getting into it. We actually had to butcher two ganders we were quite fond of because they just wouldn't stop harrassing our goat. They had their own water supply, of course, but his became a challenge. When we found they were flying over the side of his pen, we clipped back the feathers of one wing so they couldn't fly. They discovered they could squeeze through the holes in his wire fence. We ran chicken wire around the bottom. They learned to climb it. We had goose dinner and the buck finally got some peace.

In their flying days these same two

ganders got on fine with the does. The does never attempted to fight them as the buck did. They learned to stand completely still and the geese would groom them very gently. It made for an odd and entrancing sight.

Species Specifics

The Cat

Cats perform useful service . . . in combating rodents—work they necessarily perform alone and without regard to property lines.
Adlai Stevenson

Like the dog, the cat has been with us for a large part of our development. At one time the cat was used as a hunter by the human race. But when humanity began storing food, it became clear the cat was invaluable as a guardian of the stored supplies. Even today, the farm cat still stands guard over the foodstuffs, and is the best rodent control possible. That's why there are still so many cats on farms.

The cat is not a "money" animal. That is, cats don't produce income. But neither do dogs in most instances. If you have a goat to supply milk only for household use or a few chickens to produce eggs, those animals are not producing income either. Instead, they are *saving* money. It's in this category that the cat is a valuable working animal—saving you time, money and energy, and even cash in some cases.

Rodents are a serious problem on a farm or homestead. They contaminate feed, chew through binder twine on your hay bales, tunnel through everything, chew holes in your hose, carry disease to your stock, and even kill newborn animals. If you move to a new area and build from scratch, rodents from the surrounding fields move in right along with you, bringing their relatives, baggage, and pregnancies. If you buy an established place, the rodents often go with the sale. They increase in number anytime their living conditions improve, which you may do unwittingly.

You don't provide food for rodents? You carefully clean up all spilled grains? Does your horse or goat? In fact, rodents can do exceedingly well for long periods on hay chaff or grain left undigested in the manure of your animals.

Mousetraps are successful, but it takes time to locate rodents' runs, set up traps, and dispose of potentially diseased corpses. Mousetraps for mice can be used again and again; those in-the-know tell me that once you catch a rat in the trap, it doesn't work again. Live traps also can be endlessly reused, but you must drown the contents—a sordid affair for most of us.

What about cat predation on the bird population? Don't cats live on baby birds? The Audubon Society lists cats around ninth on bird population problems; air pollution and chemical spraying of fields are closer to the top of the list. Certainly cats catch birds, but far fewer than you may think. Our cats catch about two birds each a year. It hardly seems appropriate for cats to suffer persecution for two birds a year when hawks, eagles, and other birds of prey so often kill cats.

Cats preying on smaller animals tend to keep those species strong and healthy by removing the slow, sick, and weak from the population. This applies as well to the rodent population the cats guard against—they become strong and healthy living in the fields and not in the fat comfort of your house and barn.

Cats are one of the more difficult species to keep in groups. Nature disperses predator populations to the right proportion for a given area. If you can get your vet to reminisce about college days, ask him or her about the difficulties the alma mater had in maintaining a group of cats for study.

Six to eight cats are about all a small place can handle. If you have more cats than that right now, you already know they tend to come down with odd ailments. Do the best you can to exercise birth control.

There are still many farms and homesteads where withholding food from cats is still the practice. It's been long established that cats will hunt no matter how well fed. Konrad Lorenz, one of the world's leading naturalists, believes cats are *not* domestic animals and I agree. They are at best semi-domesticated and carry on as friendly wild animals. They demonstrate a remarkable capacity for loyalty to both person and place, but because of their wildness, confinement is almost intolerable for them.

We allow our cats free access to every building we have, and at the same time provide them with adequate housing. This has provided sufficient rodent control in itself. In fact, when we are having a specific problem in the area where the grain is stored, we feed the cats next to the storage. They sleep on the containers of their own volition and that is the end of the problem.

If you choose not to feed your cat population, you'll have health problems in the group. When the hunting is poor and the cats can't find anything to eat, their resistance is lowered to any disease that happens along. Disease is something you want to eliminate in all of your animals.

I know a man who had nothing but contempt for his cats. They were overpopulated, unfed and unhealthy, never wormed or even noticed a day in their lives. Naturally, they all died of distemper.*

*Cat distemper—panleukopenia—is unrelated to the canine version. It's highly contagious, and if it occurs on your premises, you must go through a period of not having cats until the incubation period of the disease is over. This period is two months long in thoroughly disinfected quarters—four months otherwise. Distemper can be completely controlled by a cheap vaccine given once a year and with a little good sense about sanitation and feeding.

The man got a few new cats, but they died, too. He tried a variety of rodent poisons with spotty success. One he tried claimed to be harmless to children and pets—a claim based on its *undesirability* to children and pets. It also said on the label (in smaller print) to keep it away from children and pets. His youngest son played with the poison, then put his hands in his mouth. While he was at the hospital with his son, his prize hunting dog ate a mouse that had died from the poison, and the dog died. These days that farmer has a higher respect for ecology—and cats.

Most homesteaders move to a rural environment desiring a healthy environment for themselves and their families. Should this be one of your reasons, remember health thrives in a healthy atmosphere and must extend to the lowest member of your expanded household. Never overlook feces tests, vaccinations, and good feeding programs for all those you share your life with. After all, it's just good sense.

CATTLE

I would rather have a ringside seat at a cattle sale than a box at the opera. . . .

E. B. WHITE,
THE POINTS OF MY COMPASS

Cattle have a rather startling history of domestication. To begin with, the majority of modern cattle breeds have forest animals as their ancestors. That is, cattle did not originate, as the American bison did, as a plains animal. But right from their start on this continent in 1519, they did very well on the open grasslands. This indicates a rather abrupt dietary change, which is uncommon in the animal kingdom. And it does explain some very basic differences between cattle and other livestock.

Cattle are extremely shy, especially so around strangers. They must always be handled with a minimum of fuss to keep from alarming them. To move cattle quickly is a contradiction in terms. This is generally true for all livestock, but even more so for cattle.

Generalities about cattle are difficult. This is further complicated by the possibility that you may have three beef animals and only one Jersey milking cow. However, some generalities apply.

All cattle need greater ventilation than is provided for in most barns, and *all* cattle, both dairy and beef, need wide doorways. Unless a doorway is four feet wide, cattle can knock out a hip joint, the same way a horse can, by banging it against a solid door frame. Be sure *all* doors for cattle open *out* and away from the animal.

Stanchions provide a way to pin a cow or a steer to a particular location when you must give it medication. With either kind of cattle, stanchions are essential but they do not have to be in the stall. If room is a problem within the stall, locate the stanchions elsewhere, but keep in mind the path cattle must take to them.

Milking is a real disaster without a stanchion. I've tried tying our dairy cow instead of using a stanchion, so I speak from experience. You sit down with your stool and bucket, cow tied short. And she

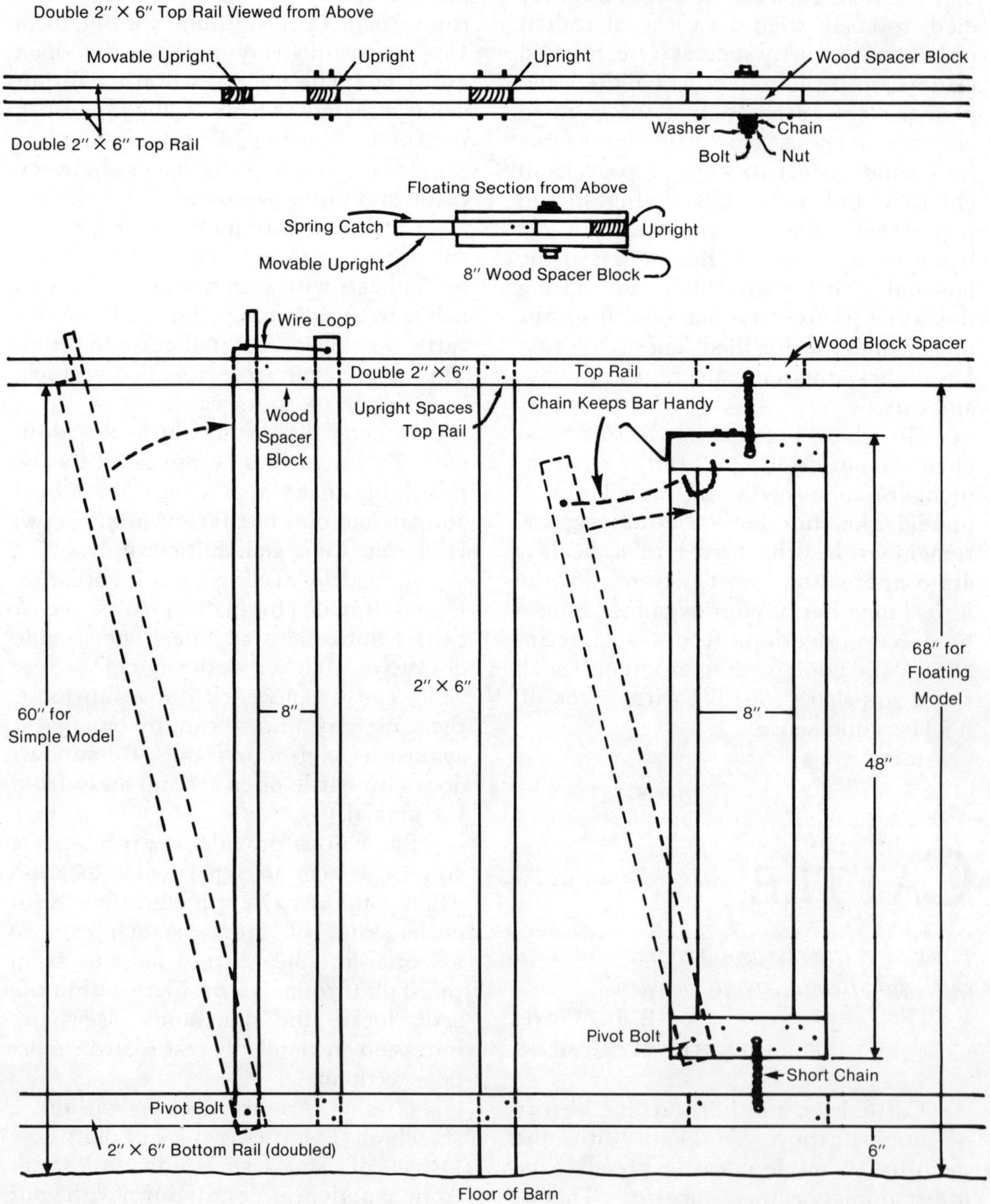

FIGURE 1. SIMPLE FLOATING COW STANCHIONS

Note: Top and bottom rails must be firmly anchored

waltzes just two steps and you, your bucket and your stool must get up and follow. As soon as you're resettled, she moves back the same two steps. It's more like dancing than milking. Tying her shorter does not help. She merely dances sideways instead. A stanchion makes life more pleasant, you a friendlier person, and your cow a pleasure to milk. It's a must.

Cattle barns tend to have low ceilings to keep initial building costs down. But such low-ceilinged barns inevitably require venting fans to exhaust moisture that the building could have been designed to reduce. To save a nickel on a length of lumber, you will continuously pay higher electric bills to run the fans. Even with a higher-ceilinged barn, if you have many cattle, you will need an exhaust fan. However you will need it less frequently, and therefore the expense of running it will be lower.

Low ceilings in cattle barns cause health problems. Damp barns kill calves, and then adult cattle, faster than anything else. If you have a problem with dampness in your barn, especially after the addition of a cow, cut a hatch above the stall or add a vent box near the cow (see Figures 65 and 69).

I highly recommend artificial insemination for breeding cattle. At least while parturition is imminent, give the cow due to calve her own stall if you have more than one cow in a stall.

The sides of cattle stalls need not be higher than 48 inches. Normally there is no need for wire above this, as most cattle are too ponderous to jump.

But the cow in the nursery rhyme—the one that "jumped over the moon"—was a heifer. Heifers are young cows and occasionally, there is one who attempts to live up to the legend. A bar across the upper part of the stall, about a foot above the partition, will stop her.

Dairy Cattle

A dairy cow needs larger, cleaner housing than a beef cow. You expect her to produce a quality product fit for your table every day, and you should expect to keep her clean enough to do so.

Dairy farmers keep cows on concrete pads in lines, one after the other, but they have to pass rigid dairy inspection rules. You, on the other hand, are attempting to keep one (or two) cows clean, not one hundred. And that's easiest in box stalls. Also if one cow falls and breaks a knee in a line of one hundred cows, there are ninety-nine left. If you have two cows, and one falls and breaks a knee, you've lost 50 percent of your "herd." Once you've worked out a cleaning system, you'll find it's easier to clean a box stall than a concrete pad full of wet, sloppy manure.

A dairy cow stall should be at least 10 feet square. If you are considering trying a cow and aren't certain you'll like having one, make the stall 12 × 12 feet, or 144 square feet. The extra space will be more flexible later, should you decide you would prefer a horse or a herd of goats or sheep in that space (see table on p. 122).

Add plenty of bedding to the dairy cow's stall and at least superficially clean the stall each day. If the housing is free, that is, she can come and go as she

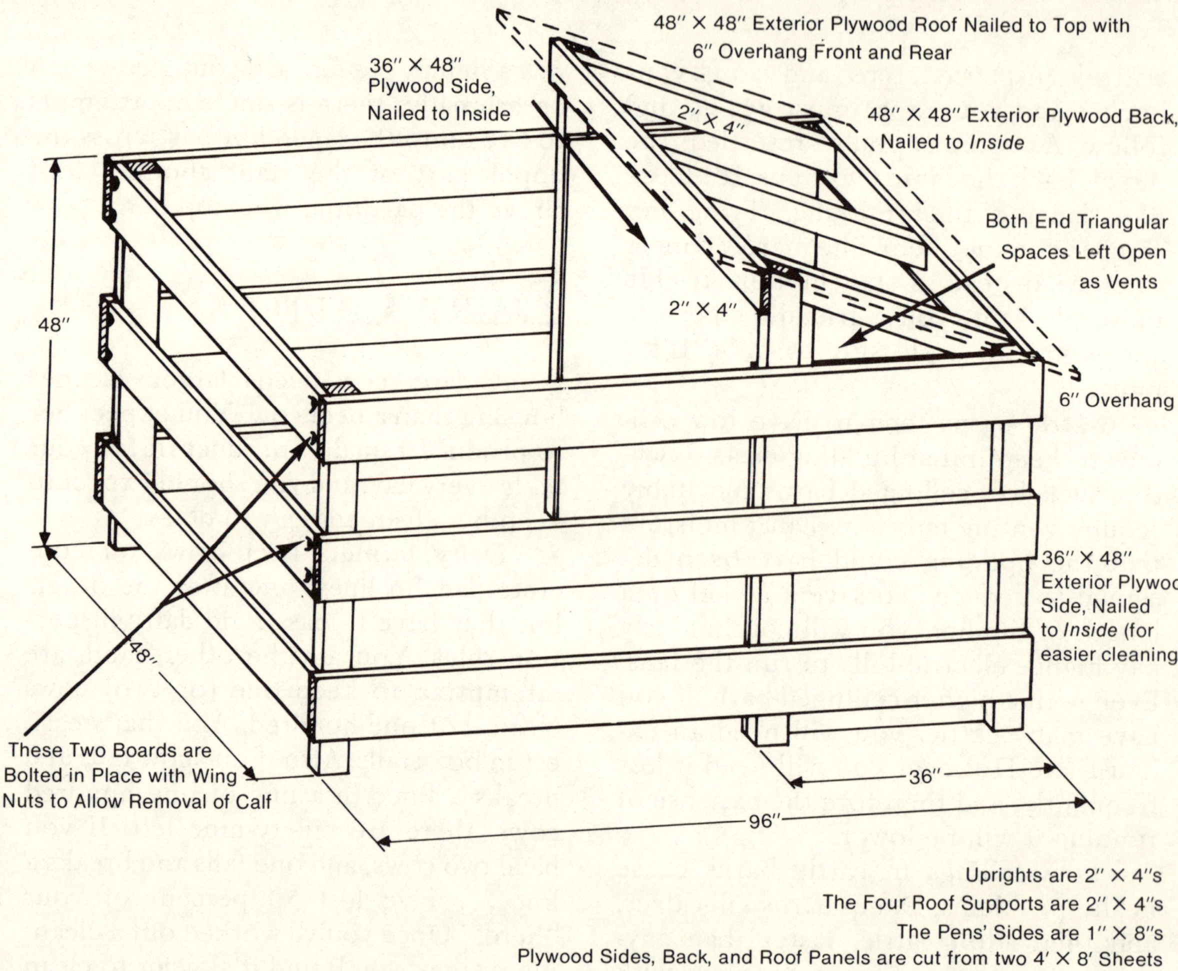

FIGURE 2. CALF PEN (One Calf Per Pen)

pleases, be certain her udder is dry after milking; frozen teats occur more easily on a damp udder.

Beef Cattle

If you have one lone steer you're raising for meat, he will enjoy a box stall as much as your dairy cow will. In fact, two steers will do well in a dairy-cow-sized stall. But with three or more, allot 60 square feet per beef cow.

Some folks keep beef cows all winter in the same stall without cleaning it. I see no harm in this as long as the stall is top-dressed as needed; that is, superficial amounts of clean bedding are sprinkled over the old bedding to keep the animals clean and dry. Clean animals bring a better price at auction, stay healthier, and are easier to work around or butcher. You may have to raise the stall sides toward the end of winter, so the cattle cannot hop over from their newly elevated positions.

Beef cattle supposedly gain weight faster when cooped up for long periods. However, healthy animals eat more and healthy animals get some exercise every day or they don't stay healthy long.

Calves

Calves do nicely in pairs in 8 × 8 stalls. Make certain neighboring animals can move out of the calves' reach. I have seen goats and hogs with no ears at all, because they were housed too close to a calf. (Calves will suck on anything stationary, and can literally exercise enough strength in sucking to eventually remove the ear or tail from a victim that cannot escape.)

Occasionally, calves get into difficulties in housing, due to bacteria. Calf housing must be cleaned thoroughly between calves. Some veterinarians recommend that calves be kept outside of the barn proper for health reasons.

Calves are cheerful creatures, and enjoy running outdoors with mad abandon. Barbed wire doesn't keep them in; sheep or turkey wire will, or even a plain board fence. As long as the pen is roomy, they'll tend to stay put.

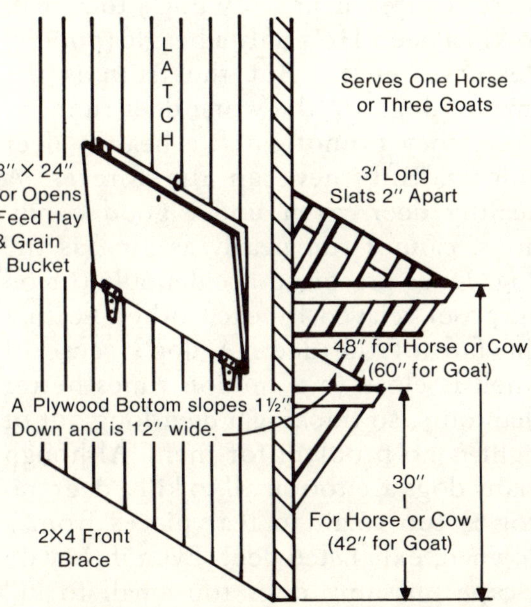

FIGURE 3. HORSE, GOAT, OR COW HAY FEEDER

The Dog

There is no faith which has never yet been broken, except that of a truly faithful dog.

KONRAD LORENZ,
KING SOLOMON'S RING

Loosely speaking, dogs and humans have done much of their evolving together. Primitive peoples with cultures based on hunting for food, probably noticed quite early how dogs could smell out and track their prey, how they barked when on the trail, and how they turned the prey back into the pack for the kill. Eventually, people and dogs apparently worked out a compatible solution to working together toward the same end: food.

As humans learned to raise animals for food, their dependence on the hunt decreased. They discovered that with certain forms of encouragement, such as breeding and better quality foods, many animals could be induced to produce more meat, milk, eggs, wool, leather, soap, and feathers. They also domesticated the dog.

The dog's capacity for loyalty made

it a natural companion. Soon s/he had developed many traits that were useful to humans. The dog gave warning when an enemy approached, helped defend the group in times of trouble, watched over livestock, caught stock for kill, provided companionship, tracked stray animals, protected the kids in a pinch, and herded animals from one place to another.

Despite many years of somewhat more tame activity, the bond between the two species continues. On any farm or homestead the stock dog is still the most valuable animal.

But there are a few problems, too. If the average dog is allowed complete freedom, s/he may develop a wanderlust and is perfectly capable of traveling twenty to forty miles a day. Dogs can do this without any particular effort. Because of their naturally active nature, dogs suffer in close confinement, unless release from that confinement is regular and dependable. Should confinement be necessary (and the nature of some dogs deems it so) arrange the housing so that the dog can see all or most of what is going on around her/him. Aside from this provision, let the dog loose under supervision at least once a day.

I have met many people who own dogs they cannot trust around their own livestock, much less around someone else's. If that's your situation, don't be embarrassed by it. The initial impetus for many people's move to the country is often ownership of a dog who needs "room to run." It's only after they make the move that they discover what their dog considers worth running after: deer, the neighbor's chickens, and their own sheep. Certainly it's sad to find this out. But it is reality. If your dog fits this category, act accordingly by restricting freedom of movement. (With some dogs, reprimands accomplish nothing.)

One thing does help: If you have two dogs, never let them loose together. Some dogs won't wander alone but two or more dogs together are ruled by pack law.

Also teach your dog to stop and sit on command. Before you let your dog wander alone, make sure you can stop her/him no matter what the circumstance. The rewards of a well-trained dog are many and worth your time: no citations for your dog running loose or, worse, shot chasing a deer; no injured stock of your own or your neighbor's; no wondering what s/he may be up to this time when s/he's missing—in brief, a little more peace in your life.

I expect a few people will be feeling defensive about letting their dog run loose. I often hear, "My dog's too small to kill a deer! He's only a beagle (poodle, Pekinese, etc.)." But the truth is the smaller dogs are the worst deer runners. True, they cannot catch a healthy deer quickly. But they can run forever. A healthy deer, even under good conditions, cannot run nearly as long as the dog. Dogs are often accidentally (or on purpose) relayed by each other, sealing the fate of the deer. A dog's sense of smell is close to a million times better than ours, so tracking a deer long out of sight is no problem for them. Although many dogs are too small to kill a deer, no dog is too small to tear pieces from a downed, exhausted deer. Even if they do escape pursuing dogs too small to kill them, many deer die each year of exhaus-

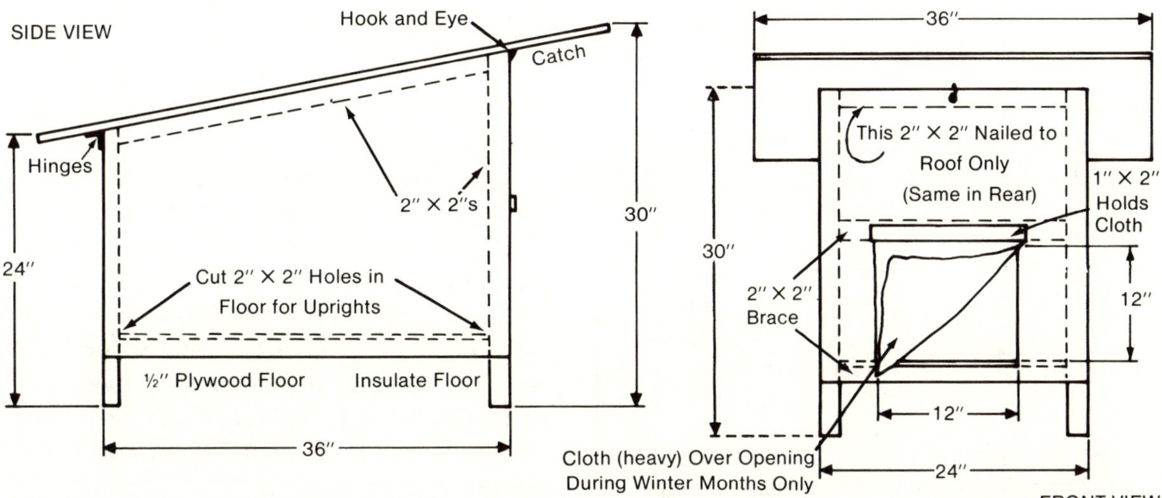

FIGURE 4. HOUSING OF 25 lb. DOG (See Bed p. 24)

tion from running too long on limited reserves.

In many locales it is now illegal to let dogs loose when there is snow on the ground. Dogs can run on top of the snow and deer fall through. On snow the slowest dog can catch the fastest deer.

The situation most to the dog's liking is to be with you always: for shopping, chores, walks, drives, washing dishes, chopping wood, everything. If you can manage this type of relationship with your dog, do! But that doesn't mean if you go somewhere and you don't plan on being gone long, your dog can't wait in the car for you! Or sleep under your bed. A well-mannered dog is welcome more places than you might think, but you might not want to put that much time into training. It's a matter of personal preference.

The most common error we make in housing dogs is to make their quarters too large. We have come to think of our dogs as family members and subconsciously expect them to want as much room as we do. But if you keep your dog outdoors, make the sleeping quarters as small as is comfortable for the size of the dog.

Consider also how often the dog will *really* use the outdoor housing. If you plan to leave the animal outdoors all winter, *don't* insulate the housing! Like livestock, dogs are not bothered by the cold but by dramatic changes in temperature (see p. 135). If you plan to use the housing only when the dog is to be left out briefly and s/he is used to being kept indoors, then do give insulation some consideration.

Peaked dog houses may look quaint but the dog would rather have a flat, slightly sloping roof. Here s/he can lie out of the mud and see what's going on in the world. When confined in a house, dogs tend to create mud—and again, you are limiting the movement of a naturally active animal. Also let the dog loose at least once a day.

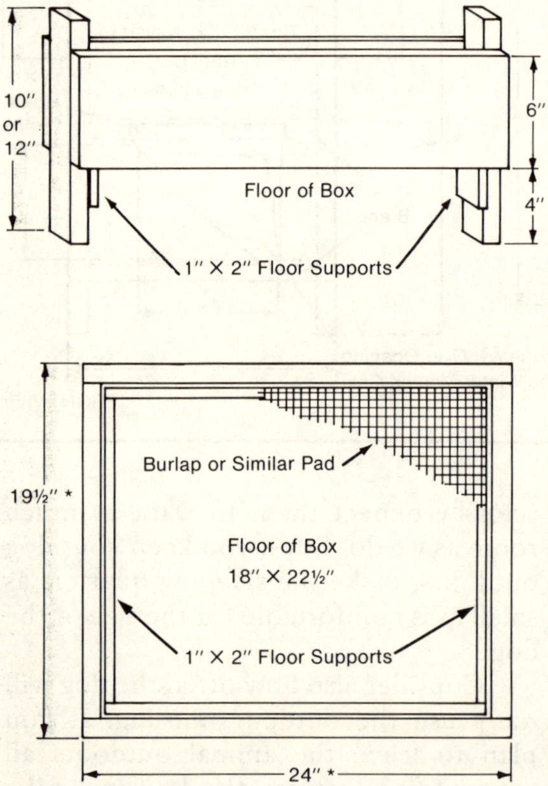

FIGURE 5. BED FOR MEDIUM-SIZE DOG*
(Approx. 25 lbs.)

*Dimensions may be determined for larger or smaller dogs by measuring the dog while curled up sleeping.

Why build dog housing at all when you can put the dog in the barn? The nature of the individual can make the difference between success and failure in the barn. The minority does well. Some dogs become bored quickly and find ways to entertain themselves not necessarily to your liking. If you have a dirt floor and/or low, open windows, your dog may dig or jump out. I've heard of several dogs that spent all their time outdoors but always managed to get back in the barn before the owner opened the door!

By nature, the dog is a predatory animal. Livestock is made up of species they would normally hunt—sheep, goats, and chickens. When you confine a dog with what s/he considers minions, there may be a serious problem. Some dogs wouldn't think of bothering livestock; others behave themselves nicely in your presence but are terrors when you aren't there.

If your dog doesn't get along with the livestock, you can bet the animals in the barn know that if the dog gets loose, they're in trouble. If your dog responds to being locked in the barn by barking, the natives of the barn will be restless. Uneasiness while eating may cause digestive upset and nervous weight loss. And what happens if Rover gets loose? Bored, the game comes naturally to such a dog: bite any nose that pokes out of any opening. If time passes and boredom grows, it's possible for the game to become much more savage without any intent on the dog's part to do harm.

What about the dog who gets along well with the stock? There is the possibility of cross-species contamination if the dog is loose. For example, a dog drinking from another species' bucket might leave tapeworm eggs on the bucket rim and thereby transfer the parasite. Play detective once in a while so you can be certain of what is going on when you aren't there. And, if it's possible, give your dog a house right in the barn—dogs like things that *belong* to them.

If you keep your dog indoors, it's a

good idea to provide a bed. It should be comfortable and the dog should know that no one will bother him/her there. When your dog gets and feels respect, s/he will behave better, just like us. And that's the beginning of a good relationship.

Other Dogs

A dog need not be wild or vicious or even brave to chase a sheep. They are only following their natural impulse to chase whatever runs. Unfortunately, sheep startle and run at the slightest disturbance. A dog too small to kill or maim a sheep can still cause death by heart failure in old ewes and abortion in pregnant ewes. Broken legs, too, often result when sheep are chased.

Paula Simmons,
Raising Sheep the Modern Way

Dogs have developed into close companions of humankind. But more and more people let their dogs run the countryside. In many communities, both rural and urban, dogs have increased to epidemic proportions. There is no doubt that dogs, singly or in packs, have become the number-one problem of the small stock raiser.

Irresponsible owners are always the cause. There is little or no problem with "wild dogs" in most locales; that rumor has been created by those who suspect their own dog is in the pack. Try to find out who owns the dog and speak to the owner. If you're told Cuddles wouldn't *do* such a thing, take a photo of Cuddles stalking your fenceline. When showing the picture to the owner, point out the sheriff might be interested.

If you can't get any cooperation from the owner, check into your state *and* local laws—occasionally they may differ. In a community where smaller-sized livestock are the meat and potatoes for a lot of people, the laws may be quite stringent. The dog laws in sheep-raising communities have the most teeth in them. In cow communities, they may be non-existent. Dogs chase and kill cows, but deer and your small livestock are easier to get. Most cows remain unbothered where the others are available.

Next bet is to watch what the dogs do coming and going. Do you also see them chasing deer? Call the game warden and have a heart-to-heart chat. They'll often help out by driving out slowly and parking obviously (so all see the car) and hunt the dogs for you. Game wardens have licenses to kill dogs running deer.

If they aren't running deer, the sheriff might help. Sometimes law enforcement officers will act as mediators in problems between dog owner and victim owner. Consider methods of trying to injure the dog(s) before resorting to your gun. I dislike the idea myself . . . but not nearly so much as I dislike finding badly torn-up defenseless animals, especially when attacked by well-fed, healthy animals out on a lark.

The gun is the last resort. Dogs can be hard to kill. And you may find yourself spending a few days in court defending your actions—days that could be better spent elsewhere. The laws usually justify your actions on the basis of "worrying or killing" livestock ("worrying" being a broad term to cover any form of harrassment), but check local and state ordinances before acting.

If you've found injured or destroyed animals and no culprit, contact the state or local law enforcement agency and file a claim against the state or county. They are responsible for the protection of you and yours, against dogs, too. If you suspect who the owner is, the courts will try to prove it and make the dog owner pay. It frequently fails and you may wind up with nothing. If you don't know who the owner is, the state or county will pay you for the loss. They pay one-fourth to one-half the animal's worth on a per-head basis. The more the government receives claims, and pays such annuities, the stricter the law will become toward irresponsible dog ownership. And *some* money for your loss is better than nothing.

Dogs rarely hesitate to stalk your stock. Sometimes it is a dog that is utterly dependable at home. In fact, the same dog may stoutly defend the same

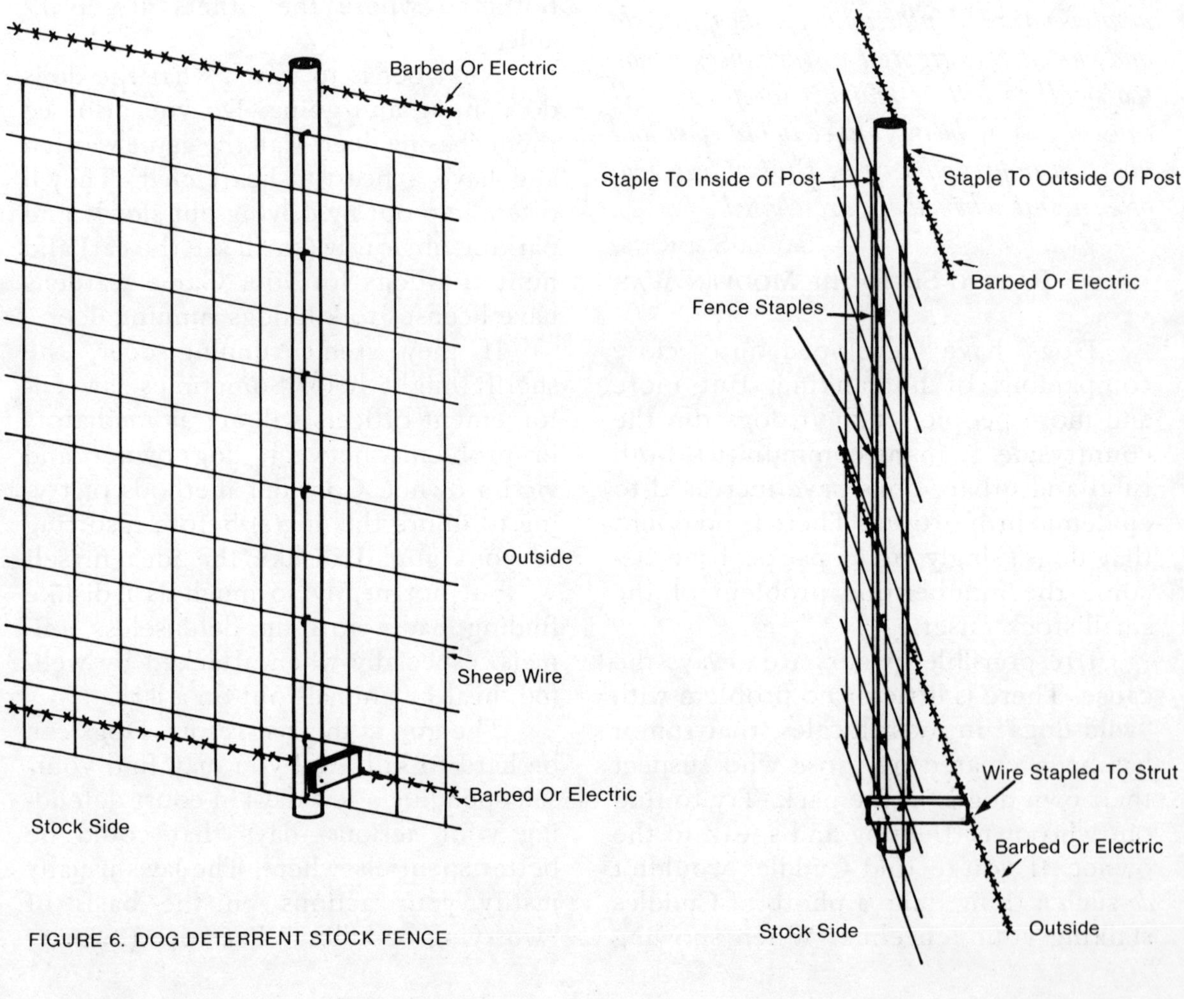

FIGURE 6. DOG DETERRENT STOCK FENCE

species at home in the light of day. It's not a contradiction to the dog; it's a way of remaining honorable at home and still satisfying the urge to hunt. It's the same type of satisfaction the trained stock dog gets from working livestock.

One last thing on the subject of "other dogs." You may have the occasional guest who shows up with a city dog they want to "let run." Be openly suspicious of the dog's behavior. Remember yourself, back in the days when you thought your city dog was only *playing* with the rabbit/kid/lamb?

But you paid for your lesson. Don't pay for theirs, too.

THE GOAT

The intelligence of goats is proverbial, many species being known for their playfulness, ingenious habits, skill in seeking protection and cleverness in obtaining food.

THE WORLD ENCYCLOPEDIA

Stock raisers are divided into two categories: those who find goats delightful, and those who cannot bear them. This section is dedicated to those of us who do enjoy goats and can tolerate their cleverness in thwarting our efforts at housing and fencing them.

The goat's natural habitat is cold, dry terrain, where food is scarce. Its reaction to a lifestyle of plenty is to waste the plenty and live on the sparse. For high quality milk and kids, the modern dairy goat needs the best food, housing, and care she can get.

Adult goats are usually easy to keep where you put them; kids are another matter. I once had two kids in a stall with six-foot walls. They used a board in the back of the stall to get themselves over the top of the partition, although it only stuck out of the wall a half-inch. I've kept others confined with no problem in a four-foot high stall with smooth walls. Smooth walls are a must considering their ability. Exercisers and feeders with adults and kids alike must be located far away from fences and walls or in the case of the feeders, built in such a way that there's no place on which to get good footing.

We run fencing to five feet in height, with a string of barbed wire along the top. Goats are not stupid and will not try to jump a fence with an obstruction above it. Kids can walk through sheep wire. Turkey wire with 2 × 4-inch spacing works better. If you must use sheep wire, run poultry mesh around the bottom on the inside.

Although there may be some disagreement, I don't think the average goat housing allows or encourages enough exercise or freedom. I've been to more than one place where the goats stand and yell all day, their manner indicating sheer frustration and boredom in a tiny "exercise" pen. Goats are intelligent and active by nature, and really suffer in close confinement. Many folks keep hay in front of them all day, and that's their prerogative. We provide enough hay at one time for them to clean it up, and encourage them to get out into the pasture whenever the weather is suitable. They're all in beautiful condition. Every one kidded well and is milk-

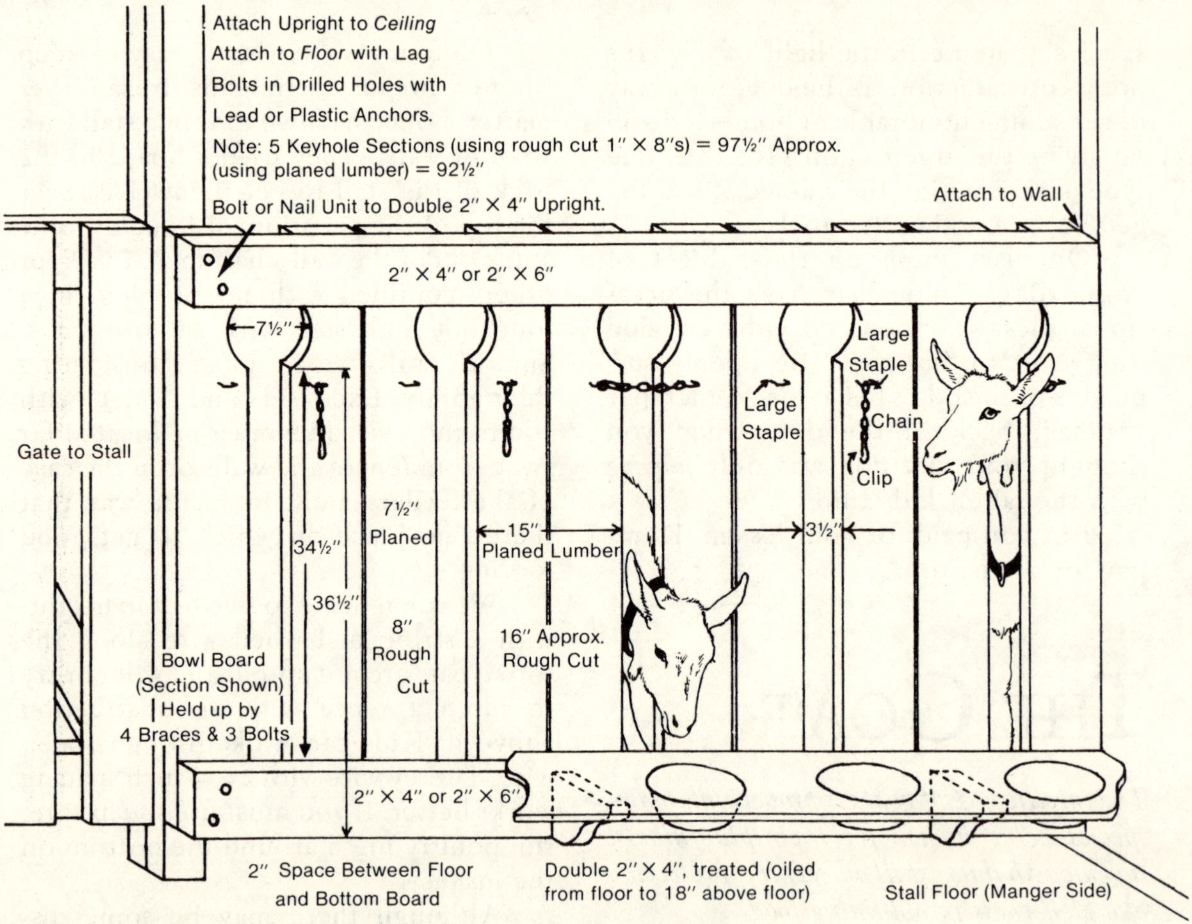

FIGURE 7. GOAT KEYHOLE MANGERS

ing better than ever. I'm surprised at how often they'll wander out and play in what I consider bad weather.

Unlike many animals, goats do not graze; they browse, eating a little here and a little there. Their pastures must be mowed periodically because goats don't eat a sufficient amount of one thing to keep fields from weeding up and getting coarse. Mowing keeps parasites down and young succulents coming up. Since goats have the largest digestive tract for their size of all the ruminants, parasites can be a serious problem, so pasture rotation is especially important with goats.

Because of their infamous ability to climb, goat mangers are traditionally built so that the goat's head can reach the food but the goat cannot climb in. Goats are extremely fussy eaters, contrary to reputation, and won't eat food that has fallen to the floor. To keep the food clean, be sure the goats cannot get into it.

Goats don't mind cold weather but they do poorly in damp quarters. The dampness holds all kinds of parasites

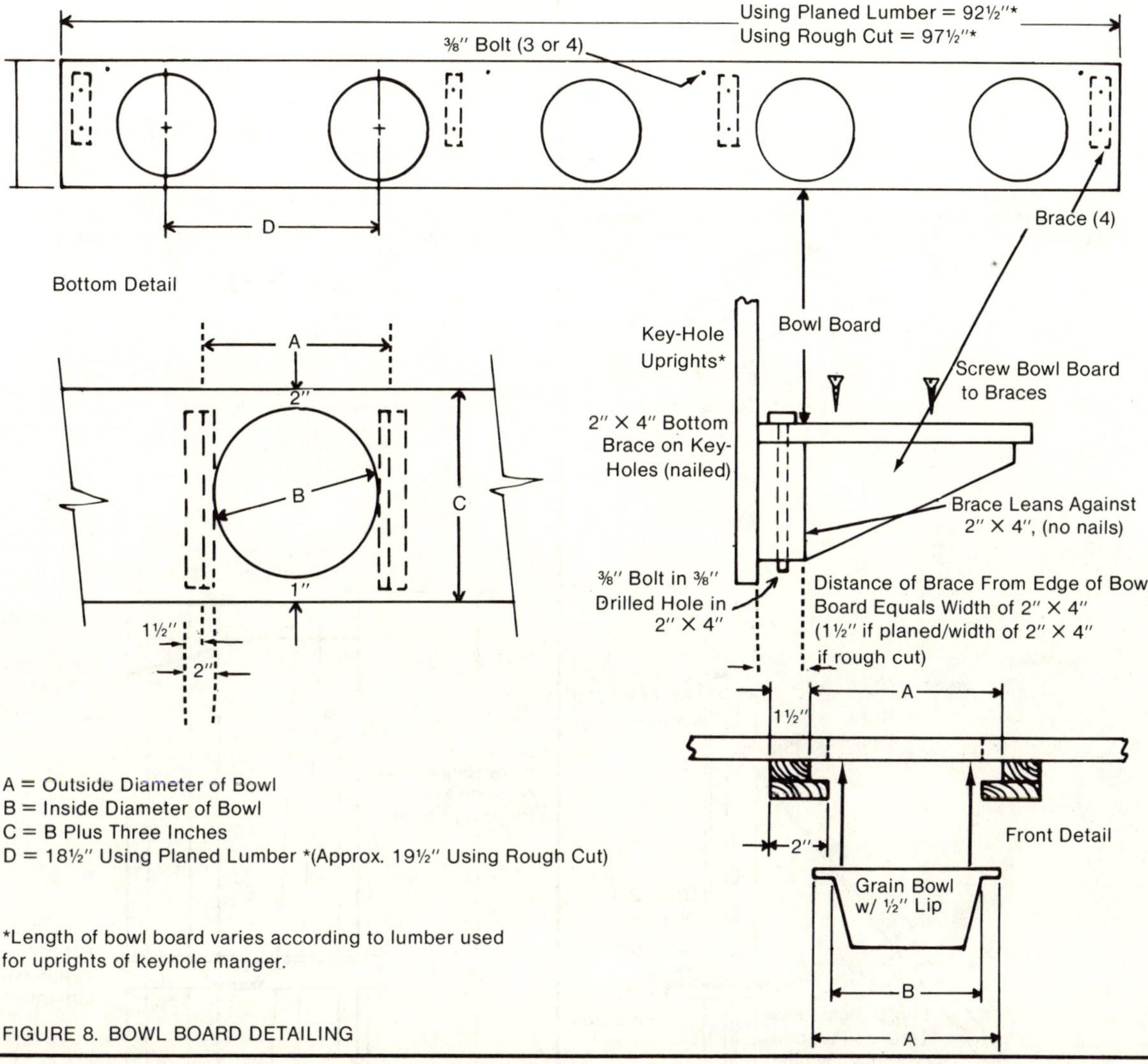

A = Outside Diameter of Bowl
B = Inside Diameter of Bowl
C = B Plus Three Inches
D = 18½″ Using Planed Lumber *(Approx. 19½″ Using Rough Cut)

*Length of bowl board varies according to lumber used for uprights of keyhole manger.

FIGURE 8. BOWL BOARD DETAILING

including stomach worms, coccidiosis, and strongyles. Keeping goats on muddy ground they can't escape is also inviting trouble; lungworm, for example, only hits goats housed on soggy ground.

Goats that are content and get plenty of exercise are usually gentle on their housing. But goats that are cooped up and bored will systematically level the barn if the construction isn't the best. They will jump on walls with their front feet, throwing their weight onto the wall with enthusiasm at the slightest provocation, so build as though you mean to keep them inside.

Bucks should have separate housing from does. For the record, it costs under $25 to have a buck descented. If he is descented, there's no reason to keep him out of the barn proper. Since your buck is half your herd genetically, he should receive high quality hay, feed, and care.

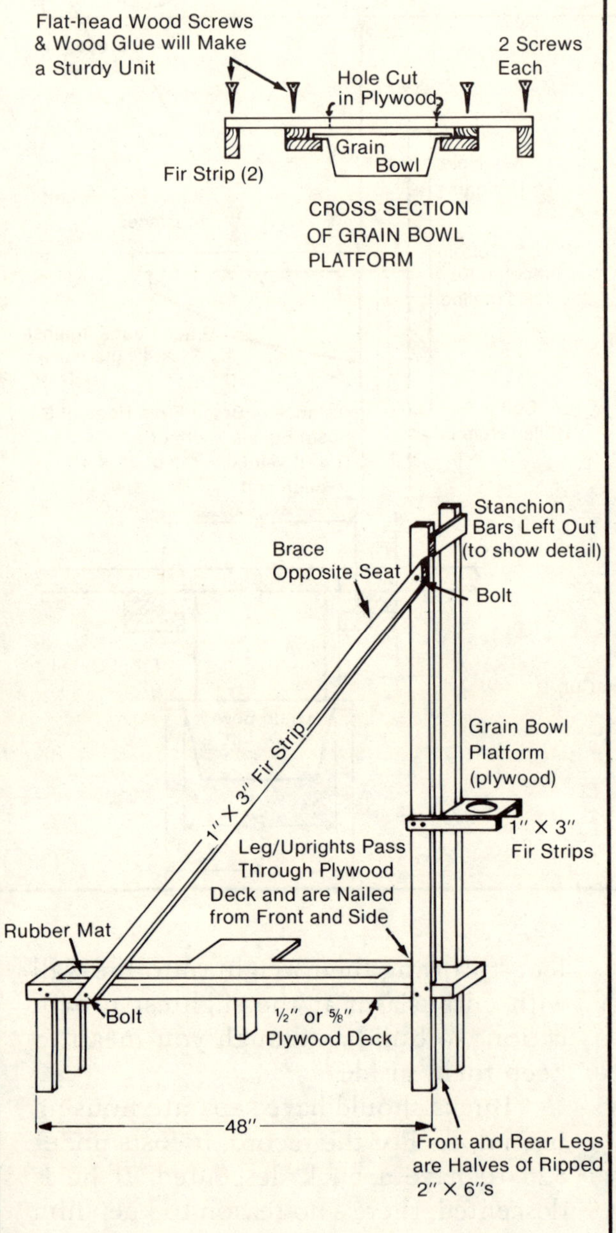

FIGURE 9. MILKING STAND FOR GOATS (SIDE VIEW)

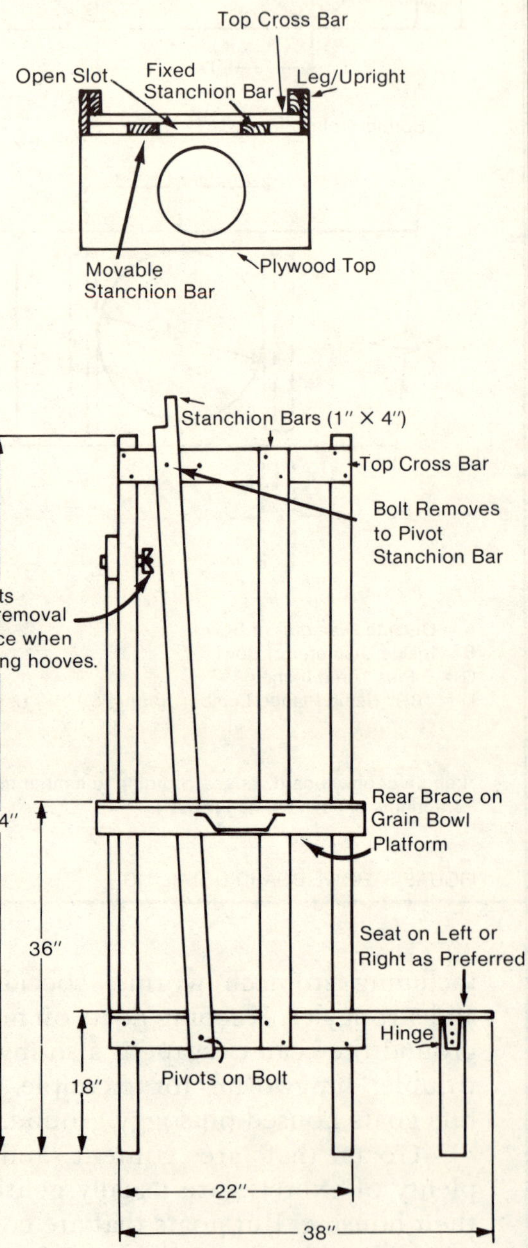

FIGURE 10. MILKING STAND FOR GOATS (FRONT VIEW)

Bucks get lonely during the off season and can be housed with horses, pigs, and sheep—*provided the individuals get along together*. They do best when housed alone, with their pen running alongside the other animals' stalls.

Bucks that are used for service outside of your own herd should be housed separately and checked regularly for any signs of illness brought in from a serviced doe. Bucks are active and intelligent, and need lots of exercise. A tire suspended where a buck can reach it will receive his devoted attention in play.

In good weather when the goats stay out all day of their own accord, we often turn the buck in with them for company. Since he knows he'll have their company again the next day, he offers no objection to being separated again at night. In the off season he has an extremely good disposition and doesn't fight with them.

The Horse

Doubtless the thorough ventilation of the frame stables and sheds is the cause of the general good health of the horses. Pure air in abundance is the one thing insisted upon for animals which may be called for at any moment to make a ride for life. . . . Open corrals or yards are usually attached to stables, and when not in use . . . the horses are turned loose. There is none of that senseless hardening of horses . . . nor on the other hand any hot-house coddling.

CAPTAIN WILLIAM E. CARTER,
Cavalry Officer, 1895

Housing has a dramatic effect on horses because of their nature and size. Keep in mind that the horse is stronger than you: she can haul you or your machinery and move faster, which makes her potentially dangerous to herself and you. Surveys of horse/people accidents indicate that more than 90 percent of such accidents could have been prevented with a little forethought. This is especially true of mishaps in the barn, where you work in close proximity with the horse day after day.

To understand the instinctive responses of horses, it is necessary to look at them under condition of their own choosing. Given freedom of choice, the horse follows a pattern similar to that of the white-tailed deer. She grazes lowlands in winter where better forage is available. Come summer, she seeks higher ground to keep cooler and to avoid the flies that hatch out of her own manure, she moves daily. With less choice available, she uses the woods to hide from the sun and at least some of the flies.

But unlike the white-tailed deer, horses are open-country animals. Because of the volume of food they must eat, and the fact that they travel in herds, they stay in large, open grassy areas. They travel a great deal to move from fly-infested areas, to seek fresh grazing, and to reach water. As a result, they have developed defenses suitable to open terrain.

Under normal circumstances, a startled or "spooked" horse depends on her speed and agility to escape danger. If speed is to work successfully, she must move instantly and at top speed. Even in uncertainty, she must run at least a few steps quickly, then turn and reappraise the situation from the slightly safer position.

This reflex creates some special problems. With their size, bulk, *and* speed, herd animals crashing into each other when alarmed can seriously injure themselves. For greater safety when grazing near each other, the larger herd animals tend to graze in the same direction. And when alarmed, they all jump in the same direction.

Kicking for the mare and striking with the forefeet for the stallion, play a role in self-defense, but I think their primary purpose is social. Rearing, however, is the only way an animal that lives by speed can pull its head and neck clear of a surprise danger in front of it. Bucking is designed for shedding a large predator before it can break the horse's spinal cord.

Understand that it's just horse sense to run from danger as the first reflex action, and that horses are consequently afraid of being caught unexpectedly. Plan your housing and handling so it will be safe for both you and your horse.

The Stall

The box stall should be a minimum of 12 by 12 feet to allow room for the horse to walk, turn, and lie down easily, and for you to enter the stall safely. A 14-by-16-foot stall would be better.

Standing stalls were created for specific purposes, such as at riding academies where the horse had to remain standing all day for quick availability. For anything but brief periods of time, standing stalls are potentially hazardous to both you and your horse. Because the horse must remain tied for long periods to prevent her from turning around, she is difficult to bed properly, and you must walk the length of the horse in order to set her free.

Partition heights between stalls should be at least 5 feet and begin 6 inches above the floor to protect the wood from rot. Since even the nicest horse will nibble on her neighbor, including other horses, it's usually necessary to either continue the board height to 7 feet above the floor or put wire mesh above the 5-foot partitions. (Make that wire a heavy gauge, chain link if possible; they'll work on it.)

Bars can be used *above* the solid wood partitions, but they have drawbacks: if they run lengthwise over the partition and are more than 6 inches apart, they'll be flexible. A bored horse poking around could get her head trapped. Poles or bars with an added support or two in the center are less dangerous, but 6-foot boards as the partition, nailed vertically for strength, are easier. The oldest and best horse barns had 8-foot partitions of solid wood between each stall; companionship to a horse is having other animals within hearing and smelling distance—they do not need to *see* them.

Make the partition between the horse and the barn proper low enough so the horse can see who is coming and going. Unlike the partition between stalls, it begins on the floor to reduce drafts. Animals that cannot see the general barn activity wear ruts around the perimeter of their stalls; this also applies

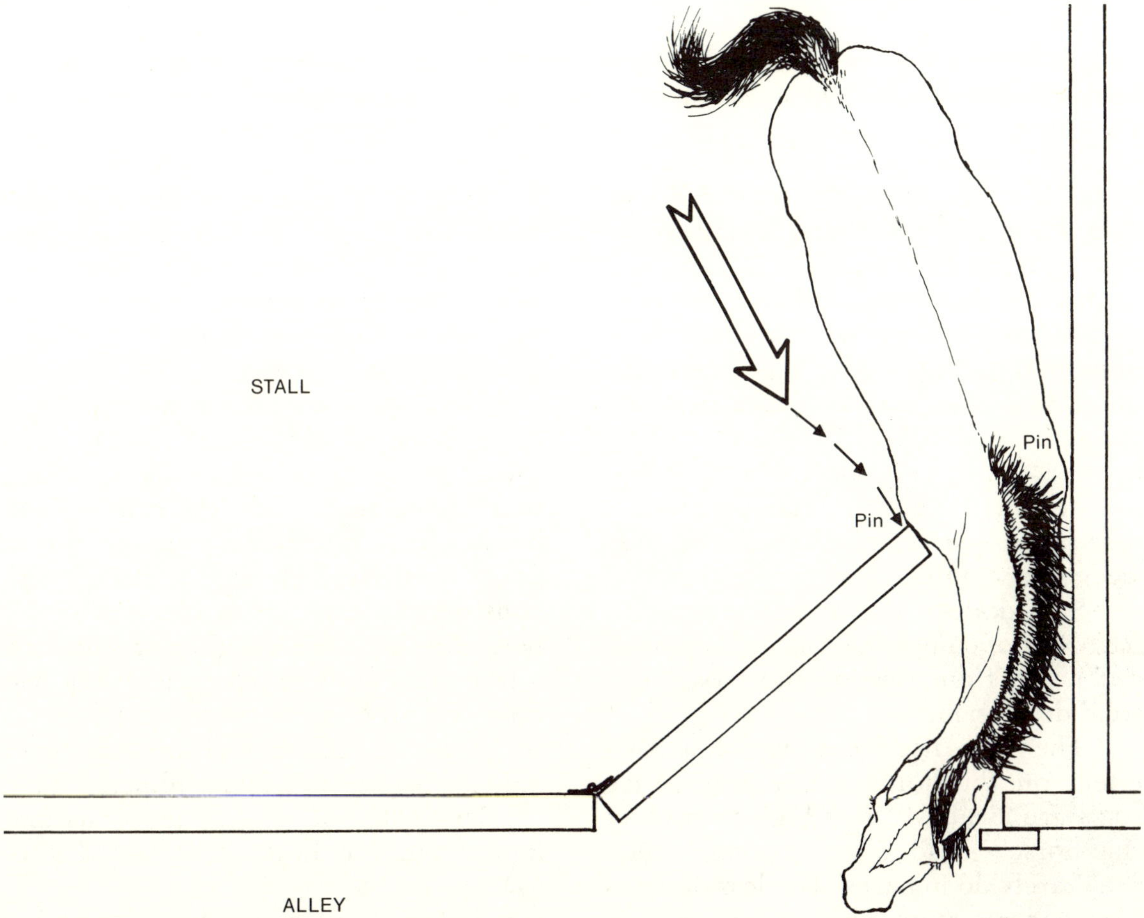

FIGURE 11. PROBLEM OF STALL DOOR OPENING *IN* TO HORSE

to the stall door, where the animal should be able to see you approach. For us, a door of 4-foot solid vertical boarding, beginning a foot above the floor and topped with 2 feet of well-supported chain-link wiring, has been the answer.

Dutch doors on horse stalls look nice. They're quite common in older barns. But a mare in season can be a real grouch and snipe at you with her teeth. On a daily basis, just trying to get by with grain for animals farther down the line can be a headache for you and the horse; she feels obligated to try for it since she can almost reach it, and you must dodge her twice a day, 365 days a year. You'll both feel more secure if she's looking through the wire behind a solid door.

The only time a stall door should be open on top is if it's located in the very end of the barn where the horse can reach nothing by leaning over it, including the door latch.

The direction a stall door opens is also important. As a rule of thumb, all external barn doors should open *out* to the outdoors; all internal doors should open *in* to the animals. The horse stall door, however, is an exception. *All* doors for the horse should open *out* away from the animal.

Backing up, except for a few clumsy steps, is *not* a natural gait for the horse. And while it's a definite advantage to teach a horse to back up, it is unnatural for it to do so. If your stall door opens in to the horse and she is at all anxious or impatient to leave, she will crowd the door. Since her head and neck fit easily through small openings, like partially opened doors, it's possible to get her pinched in the door. Under stress, she may try to rush forward, which will only wedge her tighter.

Stall doors 4-feet wide are ideal. In narrower openings the horse may knock out a hip from turning too close to a solid door frame.

The barn itself can have a strong effect on the horse. Ceilings in horse stalls are ideally 12 feet high to protect the horse's head when rearing, which they rarely do in barns. The less obvious reason is moisture control. Horses produce a good deal of moisture just in breathing, and dampness is one of the most basic health hazards to any animal. But if the majority of your animals need a lower ceiling or you already own a low-ceilinged barn, there are other options. In the loft floor, cut a hatch over the horse stall; leaving it open when necessary, the chimney effect in an otherwise well-ventilated barn will keep the air exchange sufficient to prevent moisture accumulation (see Figure 65).

Horses are more inclined to rear outside a stall. Therefore the ceiling height of the alley a horse is led through is more important for safe handling. Ideally, an alley 8 feet wide and 14 feet high is about right. If you can't come near this height, move the horse through at a reasonable speed; avoid shoeing or brushing it in an alley where a spooky horse can create problems. For alleys much less than 8 feet wide, take the horse through an outside door from the stall itself, and not through the barn proper.

No one knows your handling ability or your horses' better than you do—and those are considerations here. Many barns have somewhat low ceilings and haven't been much of a problem due to good ventilation. I have a friend who consistently leads his horses out of his barn through a narrow alley and under a 5-foot high milk pipeline and has had only one problem in twenty years. He is an expert handler, and no one ever handles his horses besides himself.

Don't forget the value of natural lighting in the horse stall (see Figure 67).

Think about how a horse scratches herself and you'll realize how dangerous stall projections are. After a tentative test, the horse rests her full weight on a wall or tree to scratch, then rubs her neck (side, rear) on the surface with great force, pushing with her legs to aid in a satisfying scratch. Now think about a bucket hook on that same wall. Or a large, splintery area.* Or a nail head. Or a

*I am referring here to splinters that form from broken, dried boards. Lumber splinters, as in green wood, are too small to bother stock through their hair coat, as they might you with bare skin. If you think about it, you'll realize that stock animals always scratch bare skin on another part of their own bodies, like noses on legs, and don't expose bare skin to small splinters.

projecting wire. Or a weak wall. All walls must be smooth and strong.

Don't leave halters on loose horses. Halters can catch on anything: the end of a gate, a board, a bucket, a hind foot. In the pasture, they're anathema to horses. Wire fencing, tree branches, and fence posts can all snag a horse halter. Besides, halters make horses only slightly easier to catch. Instead, wear a belt and loop it around her neck when you get close enough. If necessary, carry a small goody like a carrot, and you'll never have to walk back.

The cast horse is a potential housing problem. Horses can indeed sleep standing up, but they much prefer to lie down. Because of the weight involved, inertia plays a big part in the animal's ability to rise, so the process follows a set pattern: If the horse is on her side, she must roll onto her keel and lift her head first. In fact, a horse *cannot* rise unless she can pick her head up first. (This is also the reason to begin partitions no higher than 6 inches above the floor, so the horse cannot become trapped under it.)

Once her head is up, she can roll more upright and partially straighten her front legs, throwing her neck forward for balance and then swing up quickly into a full upright position. (They do *not* get to their knees first as cows and goats do.) If a horse rolls against a wall, all legs folded to the body, she cannot exert enough strength in that position to roll herself back. If she rolls against a wall in a corner, she's in double trouble; she can't roll over again and she can't get far enough away from the wall to throw her weight forward. And she will panic.

I've been in close association with horses for nearly twenty years and have only had this happen twice. It isn't that common an occurrence, but both times I needed to know what to do immediately.

Rushing thoughtlessly into such a situation is foolhardy and useless. The horse may not be able to roll back, but she can swing that heavy head and neck, and sometimes her feet, too. And no horse—*none*—is worth what you are.

Get help if at all possible. If it isn't, take your time and figure out step by step what you are going to do. Repeat your plans out loud. Your voice will calm both you and your horse, and you may "hear" a flaw in your plans.

Staying at her back, slip a rope around the base of the horse's neck. Keep both ends loose. Go as far behind her as possible and exert slow, steadily-increasing pressure to pull her away from the wall. If she's a small enough horse, you may be able to pull her free with her tail. Pull gently and increase the pressure slowly—like pulling in a fish that may or may not be fully hooked. The tail is definitely second best to using a rope because you may injure her muscles. As the horse begins to move away from the trouble spot, work more quickly if you can. She may try to regain her freedom before it's readily available.

Once freed from the wall, she may bolt to her feet. Stand clear; she'll be shaky. If she doesn't try to rise, prop up her chest on a few hay bales and let her rest. Horses are both panickers and quitters and, when on her side, she may go into a blue funk of despair. The panic may cause fluid build-up in the lungs if she's been down long. Or she may feel

FIGURE 12. PULLING CAST HORSE FROM STALL

fine and then become ill a few days later from this fluid.

When she's rested, and attempts to get up on her own, lend quiet verbal encouragement, but no physical support. She'll still be shaky, so walk her around a bit to restore her circulation. Insist gently but firmly on obedience during the process. Nothing brings an animal back to the psychological norm more quickly than the quiet firmness of proper handling.

Consult your veterinarian if the horse doesn't recuperate in a reasonable length of time.

One last and important fact about horses: they do not do well when locked up for any period of time. Both their nature and their physical well-being demand a great deal of fresh air, sunshine, and exercise. If kept in close confinement for long periods, they begin to develop physical problems. Digestive upsets and *stocking up* (where the lower legs swell from standing around too much) are the first signs that she needs more exercise. If a horse is confined much beyond the early stages of discomfort, she may develop odd and undesirable behavior. Examples of this would be wood chewing, cribbing (where she bites the crib or manger while inhaling), meanness, and weaving.

Horses love the outdoors. Keep them out as much as possible. If outdoor shelter isn't readily available in your pasture, bring her in when there are cold rains and ice storms. But, as a rule, horses like cold weather.

Ponies and donkeys do well in stalls 8 × 8 in size. We've often made a horse stall more suitable for a pony just by decreasing the height of the door so that the pony can see over it. And you can successfully keep a *friendly* pair in a 12 × 16 horse stall with no problems.

The Rabbit

Neither hares nor rabbits live in woods or forests, though they may be found on open areas within them many miles from the nearest open country.

Ivan T. Sanderson,
Living Mammals of the World

For ten years on and off, I've been raising rabbits with varying degrees of success—they can't be beaten for quality meat at low cost. I began raising them in homemade cages and hutches. (So many plans for these appear in free booklets from feed and cage companies that it is redundant to cover them here.) I then progressed to some very expensive units. When I finally achieved the "proper" housing I had been working toward, my basic dissatisfaction with traditional rabbit housing still remained unchanged.

I would like to offer an alternate solution which may give you some new ideas. Raise rabbits in a stall. Begin with five breeding does and a buck in a 6 × 8 stall. Attach wire pens with small openings from the housing for the rabbits to use in nice weather. House the buck away from the does by the simple system shown in accompanying sketches. When the does are due for breeding, remove them from their "territory" and put them in the buck's.

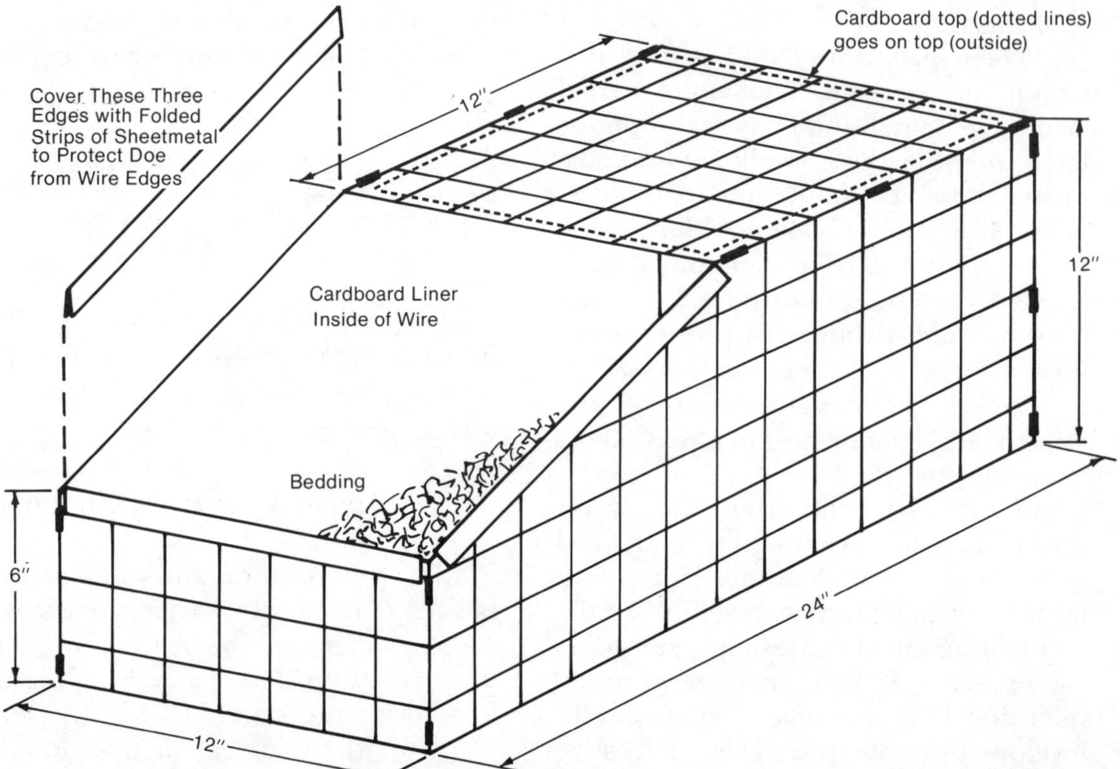

NOTE: Cardboard fits over wire on top of box to protect doe's feet when she sits up there

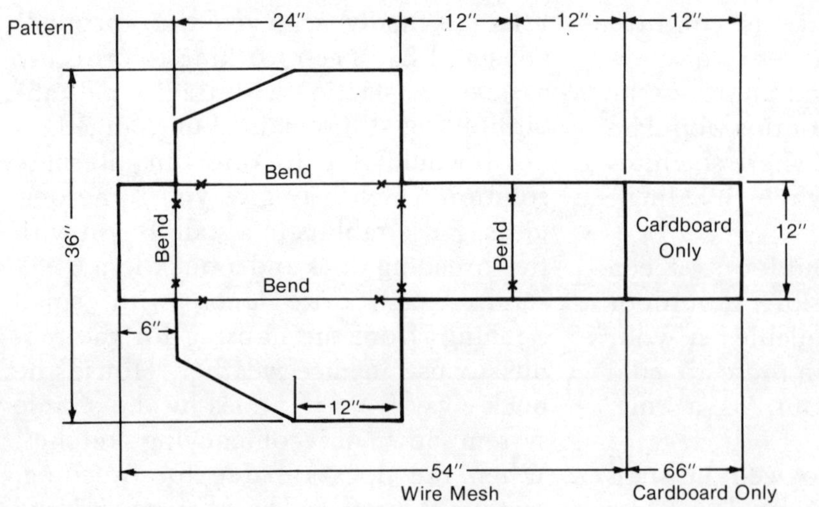

Materials
Wire mesh (1 X 2-inch squares)
3 Sheet metal strips (2 X 12 pieces)
Sheet metal clips or wire (for clipping 1 X 2 mesh together)
Cardboard

Cut the mesh and the cardboard as per pattern. Bend as shown. Clip the 1 X 2 wire mesh together as shown. Line the inside of the bottom, front, back and sides with cardboard, using short pieces of wire to hold carboard in place. Fold the three 2 X 12 inch pieces of sheet metal at one inch and place them over the wire edges as shown. Crimp in place with pliers.

After each batch of bunnies, take cardboard out and replace with fresh.

FIGURE 13. NEST BOX FOR DOE AND BUNNIES

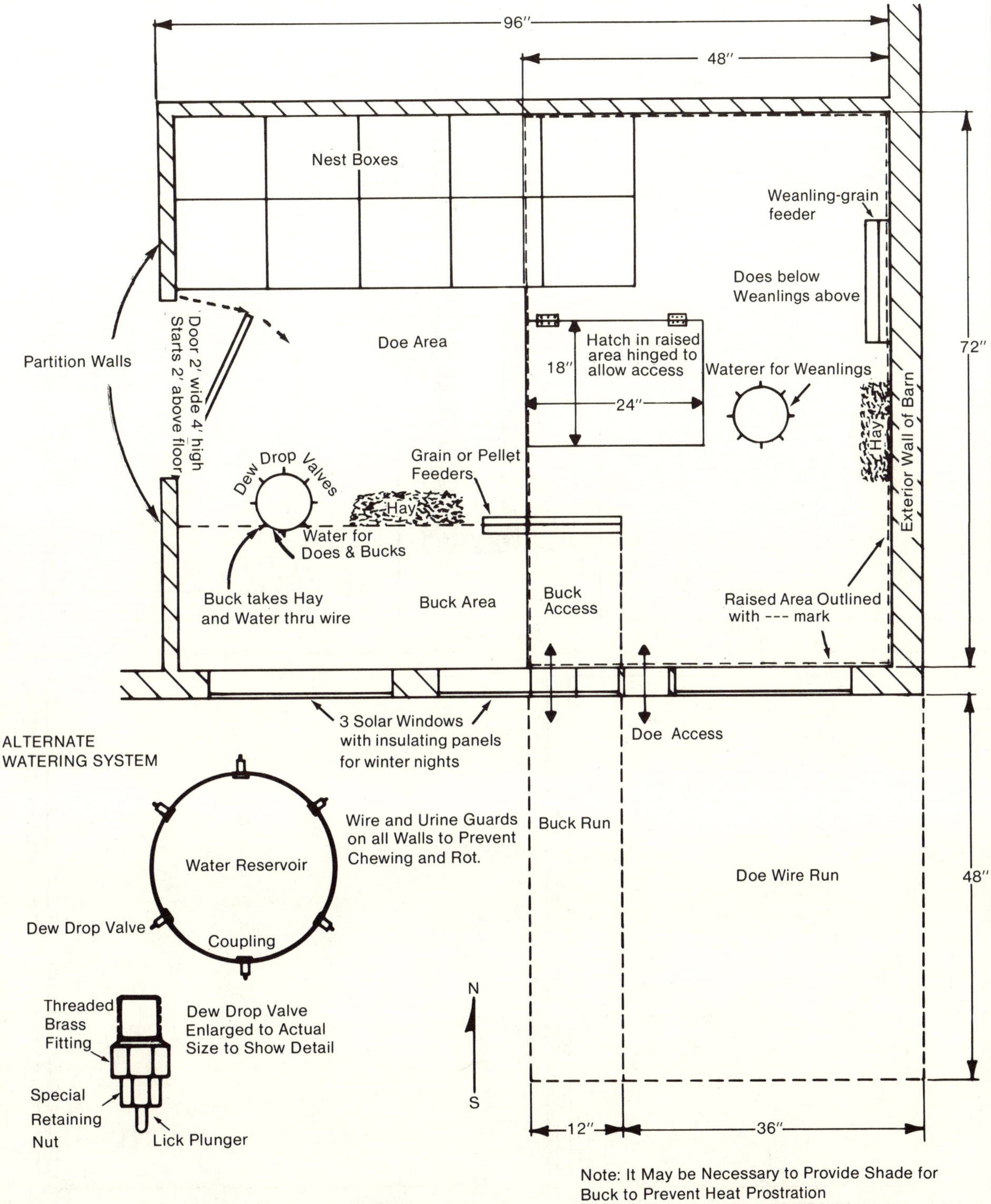

FIGURE 14. 6′ × 8′ RABBIT STALL WITH RUNS (view from above)

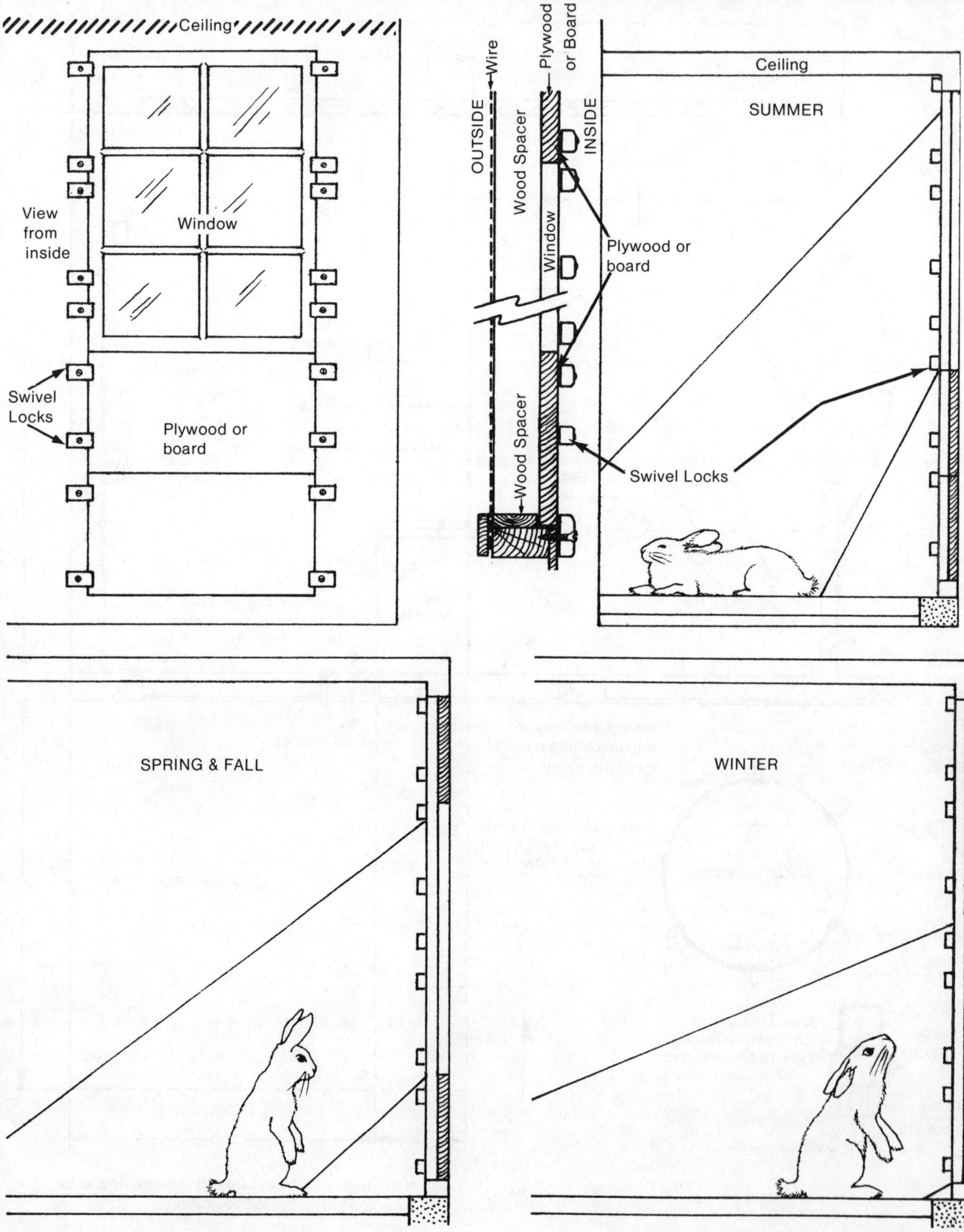

FIGURE 15. SOLAR RABBIT WINDOW

Refer frequently to Figures 14 and 17. The bottom of the hatch-type door to the stall should be a two-foot board blocking the floor level, with the door opening *in* to the rabbits. For the waterers use two-inch galvanized pipe, with short half-inch pipes out of the sides and dewdrop watering valves screwed into the ends. Insulate the tank and waterers, and protect the insulation from the rabbits by wrapping the whole unit in hardware cloth. Where the rabbits can't reach them, use heat tapes to keep the units flowing in the worst weather. Arrange feeders and waterers for the basic breeding stock so that they can be filled from outside the stall.

Locate nest boxes in a line where they will receive the maximum sunlight from the solar rabbit window. Above them build a small platform for a single, insulating layer of hay bales in winter. Of course, you won't need them in there all the time or all at once if you rotate the does' breeding.

The insulating panels for the solar window can be made of any insulating material enclosed in plywood. They should be moveable and removeable depending on the season of the year.

Keep track of the production of individual does by their ear tattoos. For easier daily identification, also mark the ears with an animal marker pen. On a nail by the door keep a chart with the does' tattoo numbers and corresponding ear marks for checks on past and present production.

Concrete flooring is a must in this arrangement to prevent the rabbits from

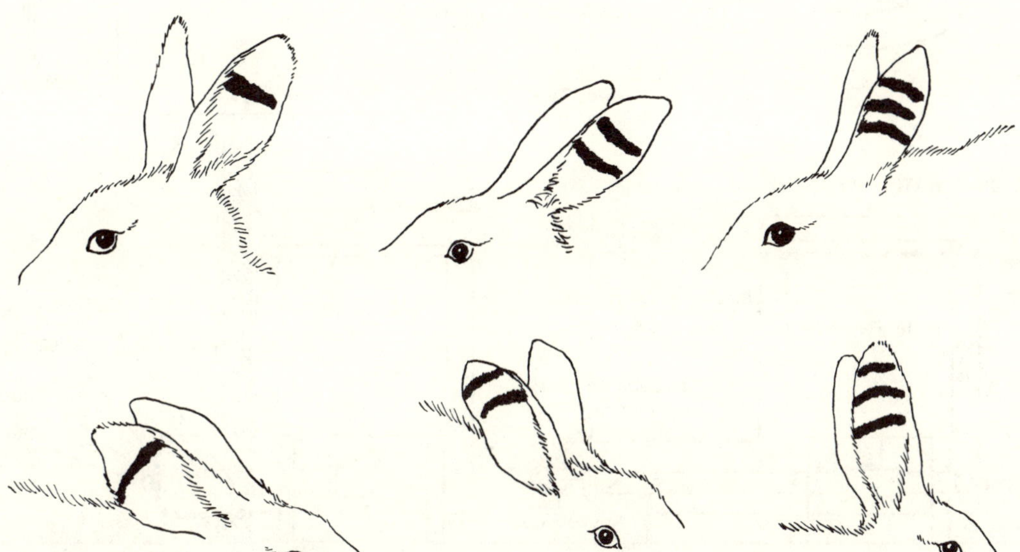

FIGURE 16. EAR MARKING SYSTEM

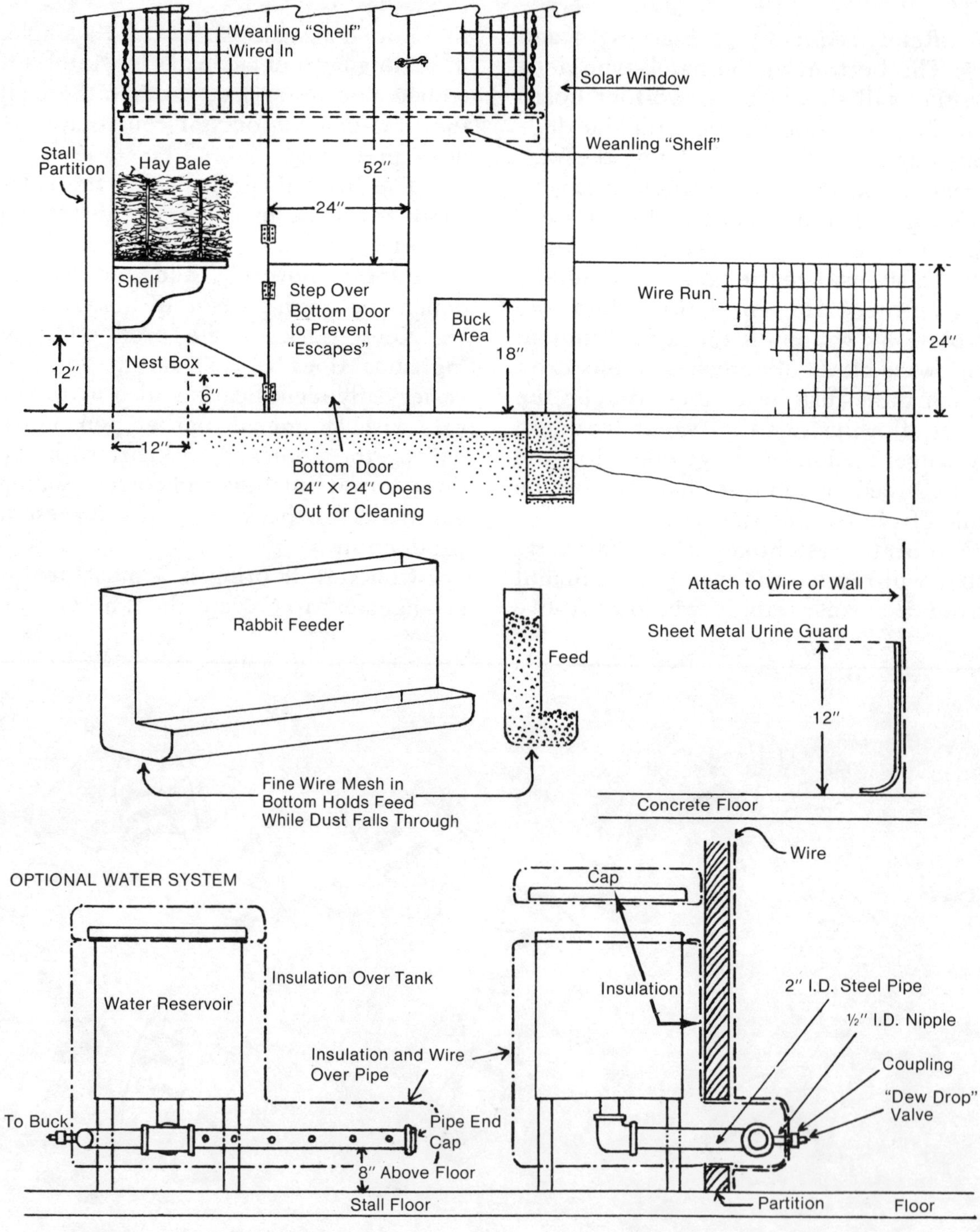

FIGURE 17. RABBIT STALL (side view)

digging their way out. For the same reason, as well as for wall protection, install flashing around the base of the exterior stall walls on the rabbits' side; it will protect the gap between walls and concrete floor. Sheet metal flashing is sold in three-foot-wide strips. Cut in half lengthwise. Cut again into lengths to fit the dimensions of the stall. Bend a four-inch curve lengthwise into each piece. Place the edge on the floor; attach the twelve-inch width to the walls. On top of the floor and flashing, place bedding to protect the rabbits' feet and to make regular stall cleaning easier.

House weanlings until butchering on the "upstairs" floor, located on the wall opposite the stall door. Build this suspended floor from a 4 × 6 sheet of plywood, rigidified with 2 × 4s nailed to the bottom. Treat the top surface of the plywood with tar and add bedding. Suspend the floor from the ceiling with chains. It will fill the end of the stall, with the walls as end-closures. Put wire over the walls to protect them from being chewed. Cut out a two-foot square from the center and hinge it to the platform so that it will swing down. Again, reinforce all edges with 2 × 4s. Add a latch so that you can open and shut the hatch. Whenever you need to enter the cage, open the trap door slowly—the young rabbits will hop off. Smaller openings in the wire front are practical for feeding and watering.

This design for rabbit housing costs the same as traditional cages, feeders, and waterers for three does and a buck.

Maintain one or two traditional cages outside the stall area for newcomers and quarantined rabbits, but I don't think the conventional cage system is all it's cracked up to be. While the cage arrangement seems to make work as easy as possible, I'm not convinced it actually succeeds—except on a commercial scale. But cages may suit your needs better if you're cramped for space.

SHEEP

Modern sheep raising has shown a real trend toward the small situation, with emphasis on profitable self-sufficiency.

PAULA SIMMONS,
RAISING SHEEP THE MODERN WAY

Sheep have been domesticated for so long that little is known of their ancestors in the wild. Certainly, there are wild sheep, but the actual ancestry is more confused than that. Among other problems, they have many mutual ancestors with goats. It is safe to say that sheep, in their modern form, are the product of thousands of generations of selective breeding.

Sheep and goat housing differ because of differences in behavior between the two species. Goats are highly individualistic and downright rowdy; sheep are quieter, and more concerned with the groups' activity than goats are.

For years, I thought sheep were just stupid, unlike goats. After I acquired a few, I changed my mind. Sheep simply prefer a peaceful atmosphere and group interaction more than goats do.

Like any other animal, sheep will

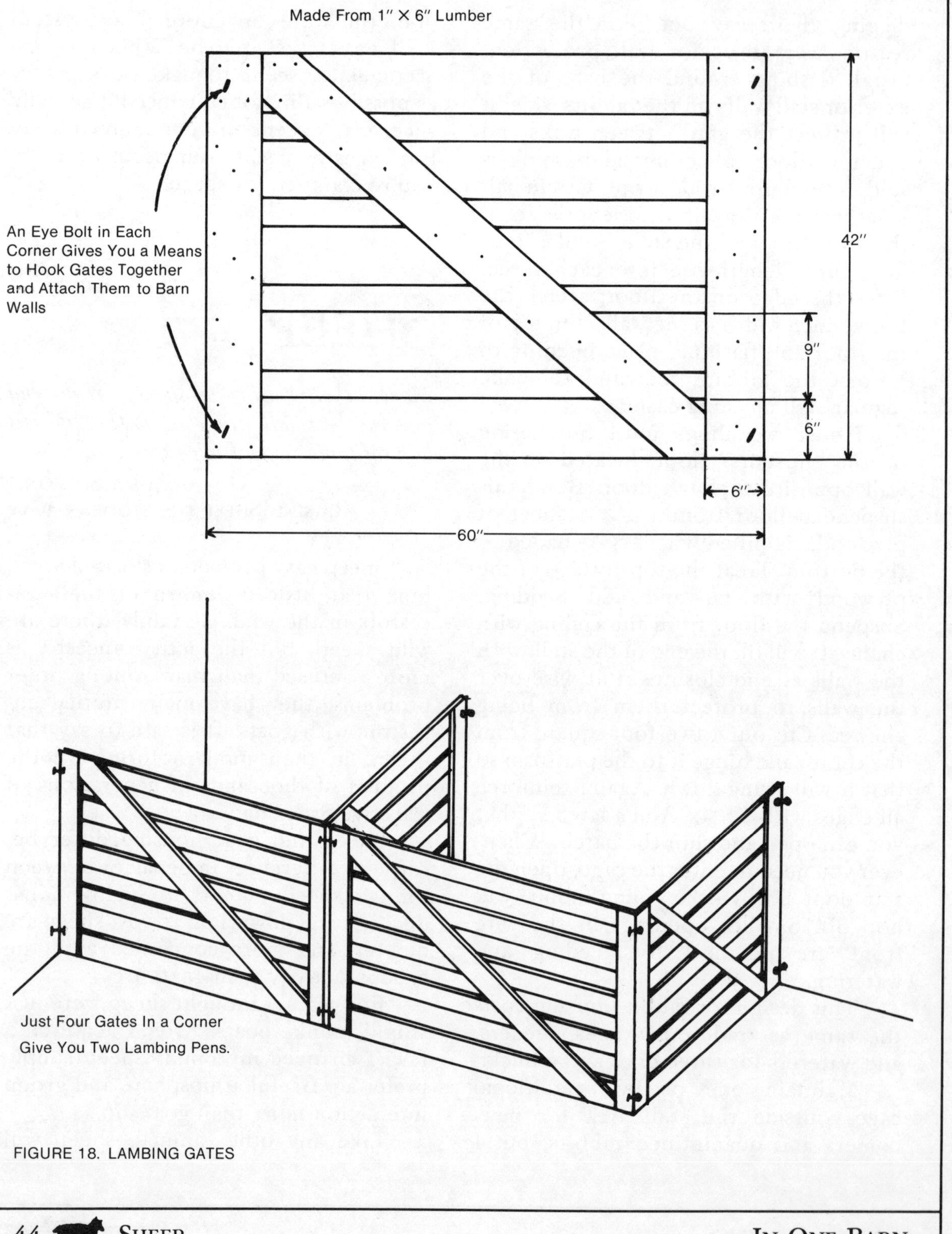

FIGURE 18. LAMBING GATES

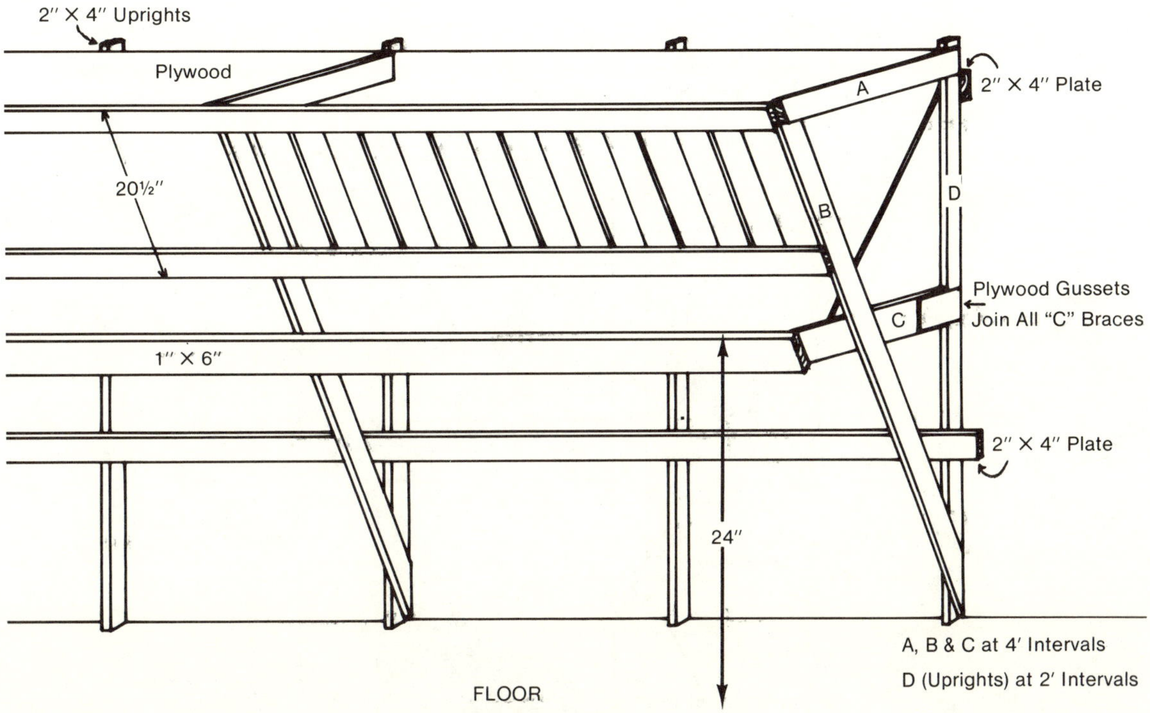

FIGURE 19. HAY, GRAIN AND ALFALFA CUBE FEEDER FOR SHEEP

constantly try to break out of inadequate quarters or pasture. But in my experience, sheep are easier to fence than goats are. I have seen many a sheep farm with fences that a goat would step over, and the farmer had no trouble keeping them in.

The best fencing for sheep is sheep wire. Running a hot electric wire along the bottom keeps out dogs. Even a single strand of barbed wire slows the average dog down considerably. Since a ewe's wool insulates her from a shock from an electric fence, she must be trained to stay inside it. To accomplish this, hang bits of wire from the electric fencing. A curious ewe will sniff it; once she finds out it bites, she'll stay away from it. Three strands of electric wire are enough to make it difficult for a ewe or lamb to get out or a stray dog to get in.

Dogs are an ever-present danger with sheep. Bring in your sheep at night if dogs are a serious problem in your area. Stock people everywhere are trying to solve the problems between sheep and dogs. Donkeys, llamas, even livestock guarding dogs are being put in with sheep. If you are having problems, check with other sheepraisers. See what is being done in your area. The important point is that you're not alone.

Sheep must have sufficient graze. Pasture rotation is also very important with sheep to prevent overgrazing, parasite problems, and soil erosion from ewes using the same pathways every day.

Higher, drier ground is better for

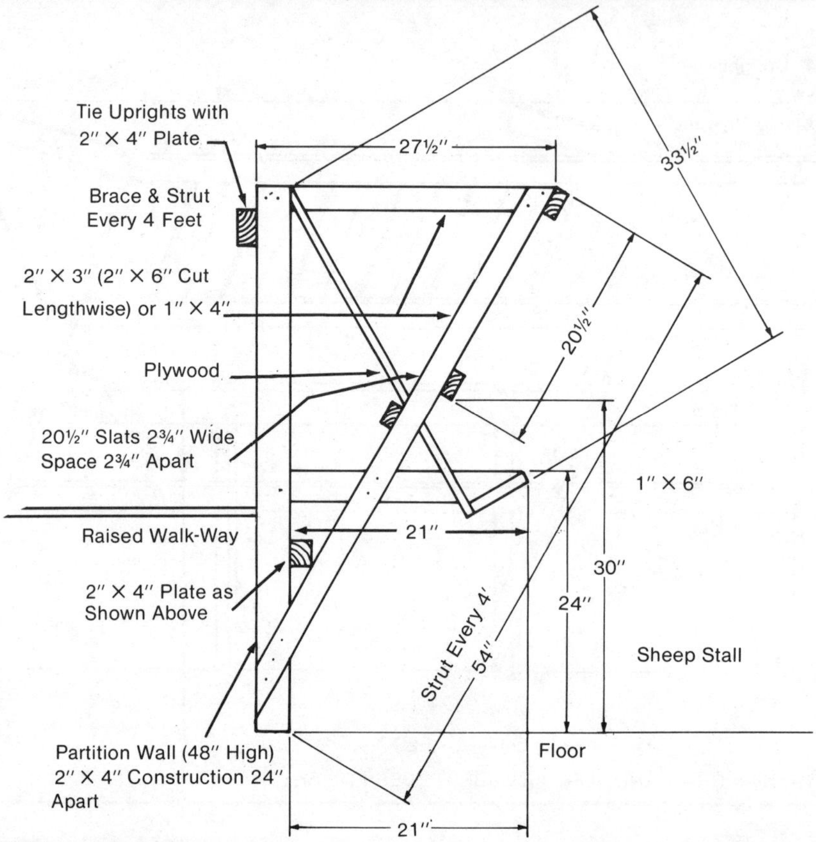

FIGURE 20. SHEEP FEEDER (cross-section)

sheep, since foot infections are a serious problem with this species. And shade in summer is a must.

Since sheep take an active part in group life, they move, eat, and are housed as a unit. Allow twenty square feet of housing per ewe. If sheep are overcrowded, problems will develop in the group; you can settle those by splitting the group or enlarging the quarters. Doors for sheep should be wide enough for them to move through comfortably in twos and threes.

Lambing is the only time sheep need more than a minimum of housing. Use portable gates (see Figure 18) to separate ewes that are lambing from the group and from each other. They can still see and interact somewhat with the rest of the herd, so there's less stress than isolation would cause. After birth keep the new lambs in a pen with their dams for a few days until the bonds between them are strong. This will also encourage the ewe's mothering instincts.

After breeding season (September through December) give the ram separate quarters. Either keep him in with the buck goat, or share a ram with a neighbor. The ram does not have to be kept isolated, but should be separated from young ewes if you do not want them bred. Remember to take good care of him, even in the off season.

Keyhole feeders are unnecessary for sheep; they don't waste food the way goats do.

Bottle lambs do well housed with kids. I frankly don't know if you can get a lamb to drink successfully from a pan as you can a kid. Lambs aren't as talented as early as kids, but they are incredibly lovable.

SWINE

The industrialization of animal husbandry is . . . seriously oversimplified. In addition to the ethical questions involved, the use of animals as machines—penning them in feed lots and cages—creates an enormous pollution problem.

WENDELL BERRY
THE UNSETTLING OF AMERICA

PIGS

Although it surprises many people, the pig is basically a woodland creature. Pigs eat all forms of vegetation but survive mainly on roots, snails, worms and other edibles found by digging and rooting. They, like us, are omnivores and can do very well on our diet. In one way, it may be the demise of the pig as a domestic animal, since pigs can compete directly with humanity for food. This will become an important factor as the world food supply continues to shrink.

Given the chance, pigs are the tidiest animals in the barnyard. The only time pigs are the least sloppy is in eating, when their enthusiasm and greed get the better of them. But they are the only barnyard animal that won't overeat.

Many of the pigs' health problems come from damp quarters. Their heads are so low that damp floors or bedding cause them to breathe in moisture all day. Pigs have dry skin and enjoy soaking in mud wallows for their coolness and moisture. Kept in small quarters on mud floors, pigs will make new wallows as soon as the old ones become dirty. Very quickly, there is nowhere to make a new wallow. In domestication, the pig's fabulous ability to dig is a problem. In the mud of the pen, parasites are carried from one generation to the next. The solution is to change pens or to keep the pigs on concrete. We prefer the concrete; it's easier to keep clean and dry.

Because of their impressive strength, pigs require well-built pens. One-inch-wide boards or half-inch plywood is inappropriate unless it's well supported and nailed securely to two-by lumber (2 × 4, 2 × 6, etc.). Use at least 16d nails. Because few pigs are able to climb or jump with agility, height on pig structures is adequate at forty inches (see Figures 26 and 27).

Pigs will eat good quality hay, although in my experience, not much of it. I provide some hay on a raised pallet they use as a sleeping place; they love to root through it when lying down. I tried a hay feeder; they pulled it out and rooted it all over the floor. Since their floors should be kept dry, don't use any bedding. It will hold moisture. I feed all the grains wet in the feeder shown in Figure 21.

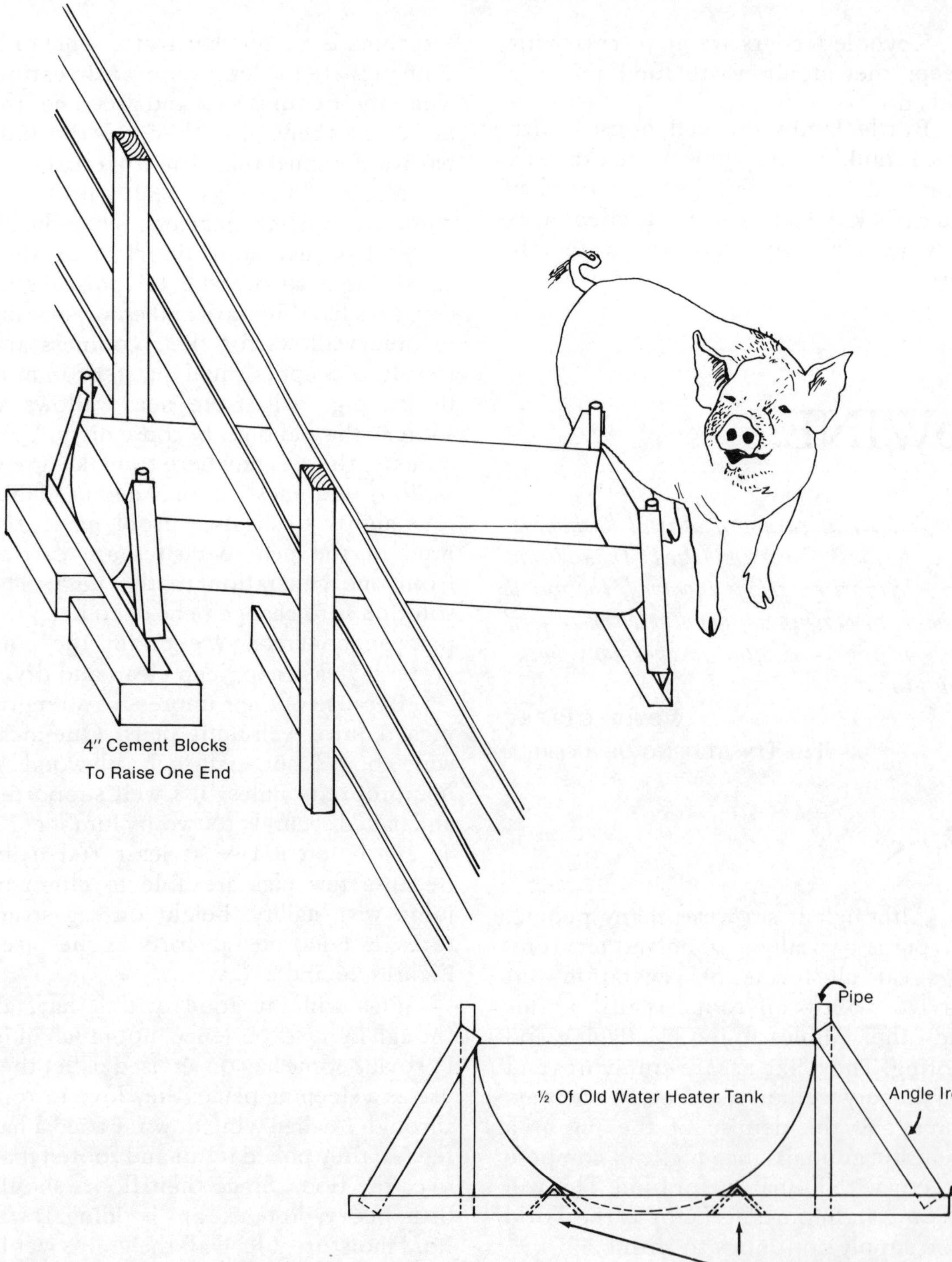

FIGURE 21. FEEDING TROUGH

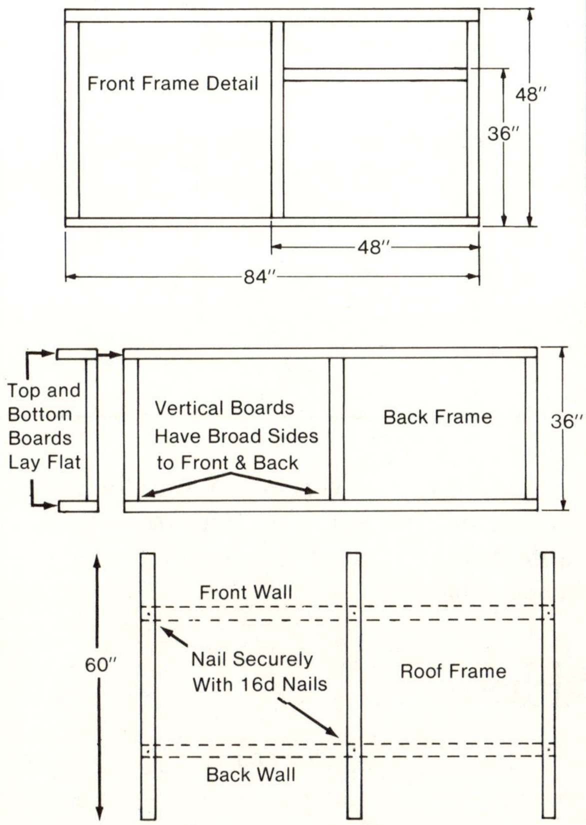

FIGURE 22. PIG OR HOG SHED (Holds 2 Adults)

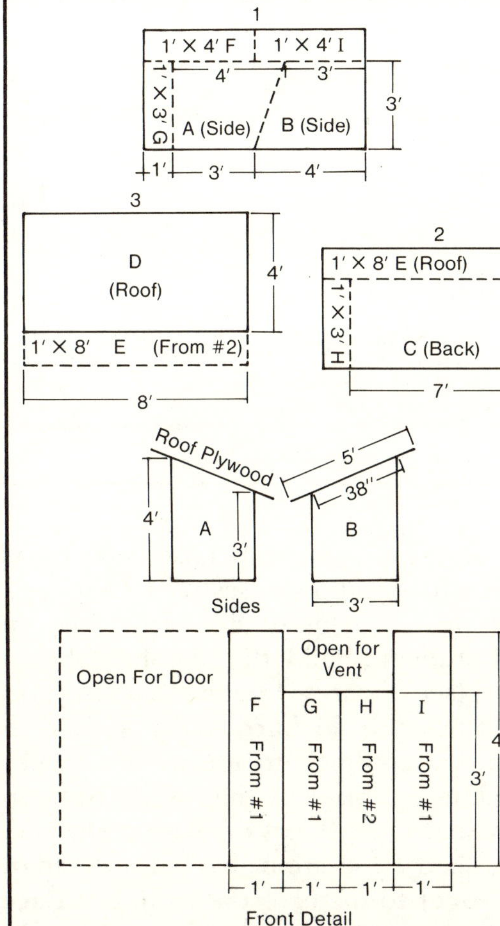

Face Door Opening to South (or East)
FIGURE 23. PIG OR HOG SHED (Sheathing Detail)

When fed grains mixed with water, my pigs grow a third again as well. Above and beyond that, I supply fresh water twice a day in a well-fastened rubber bucket.

The pen layout that seems most satisfactory to myself and the pigs has separate areas for eating, sleeping, and "bathroom" activities. I locate the sleeping pallet where the pigs can see or hear me coming, but far enough away from the eating area so they won't mess it up in the excitement of dinnertime. Pigs seem to prefer a raised sleeping area except when it's hot; then they stretch out on the bare concrete alongside it. I also put sawdust in a corner of the pen as far away from sleeping and eating areas as possible. Pigs that haven't already acclimated to soiled quarters will use this area exclusively for a "toilet." Sawdust works best because they find little in it to root for.

Pigs will go outdoors as often as they can if such an arrangement is provided. They will graze right along with other stock. If there isn't sufficient ground for the number of pigs you put out, the area will look like a battlefield in no time. Excellent fencing is a must.

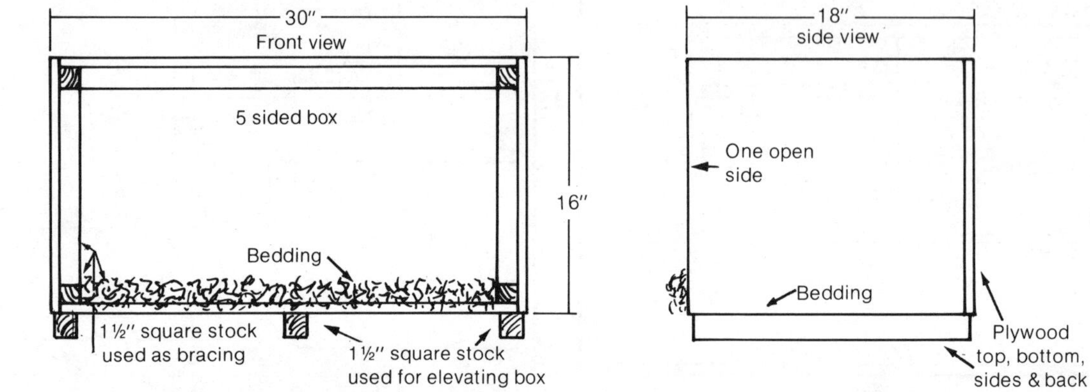

FIGURE 24. DETAILS OF PIGLET BOX

Right now, we only raise two pigs at a time for meat purposes. Years ago, as a class project in building construction, we had students build the pig shed shown here (see Figures 22 and 23). This was built prior to the barn and it worked so well, we continue to use it. It's a light shed held in place by the 7 × 10-foot pen attached to it. It sits on a concrete pad. The shed 2 × 4s are nailed flat (instead of on edge) to increase the nailing surface and to better support the plywood sheathing. The shed is liberally coated with drainage oil as a preservative on the outside, whitewashed inside. And while we climb over the side of the attached pen when we need to get in and take the pen side down to remove a pig, gates are more appropriate in the barn.

Make gates exceedingly stout. If you wish a swinging gate it should open *in* to the pigs and close *onto* the bracing of the pen's side for added support; otherwise, gates should be moveable types, taken off completely on the few occasions when a hog is to be moved from one place to another, as in the case of the farrowing pen gate.

If you've never raised pigs before, you're in for a surprise. Pigs are personable, loyal to each other, intelligent, and affectionate.

Hogs

In his book, *The Homesteader's Handbook to Raising Small Livestock*, Jerry Belanger says: "The average homesteader would do well to forget about a sow . . . when all you want is meat for your table." And he is right. It's very unlikely that the average small stock holder is going to raise hogs. Instead, such a stock holder is going to raise pigs, the younger and smaller version of hogs.

However, if you get into hog housing in a serious way, give each sow her own regular stall, with a separate farrowing stall. Further, a boar should have his own stall, and the breeding should take place in the pasture.

If you get into hogs in a commercial way, house the boar separately so he won't bring in potential diseases from outside-serviced sows to your own sows.

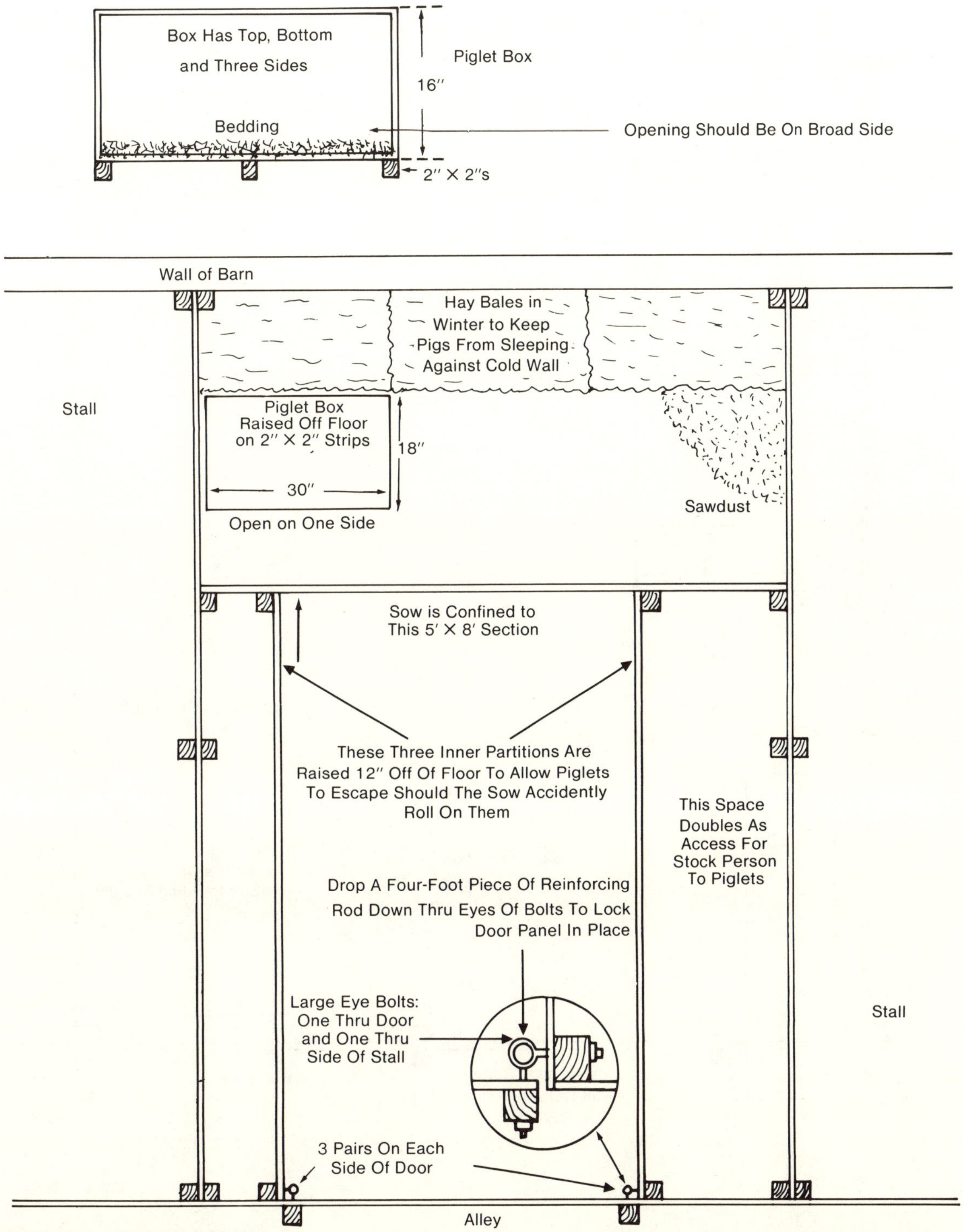

FIGURE 25. FARROWING STALL

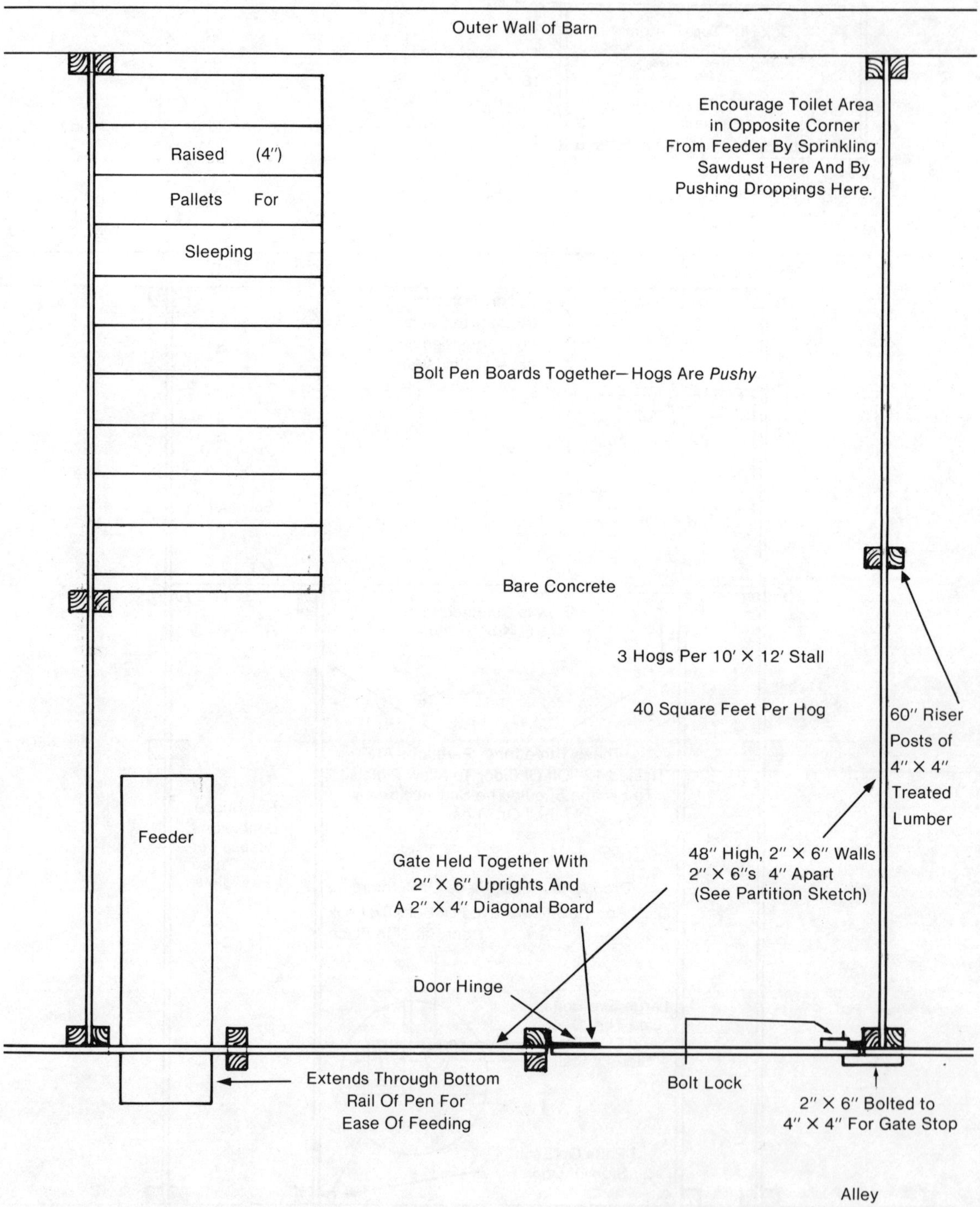

FIGURE 26. PIG OR HOG STALL

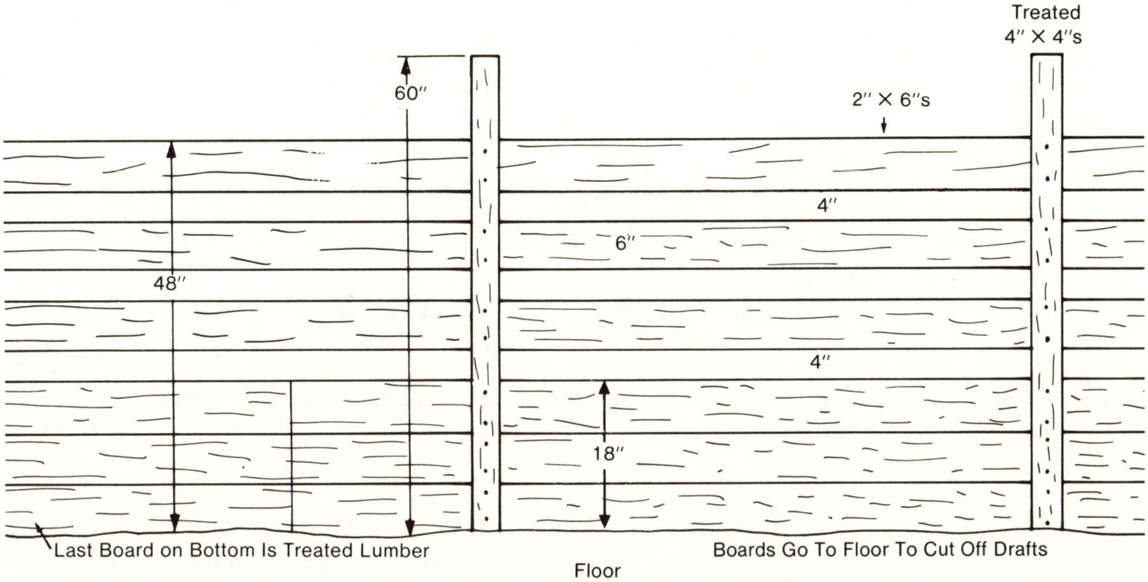

PARTITIONS FOR PIG OR HOG STALLS

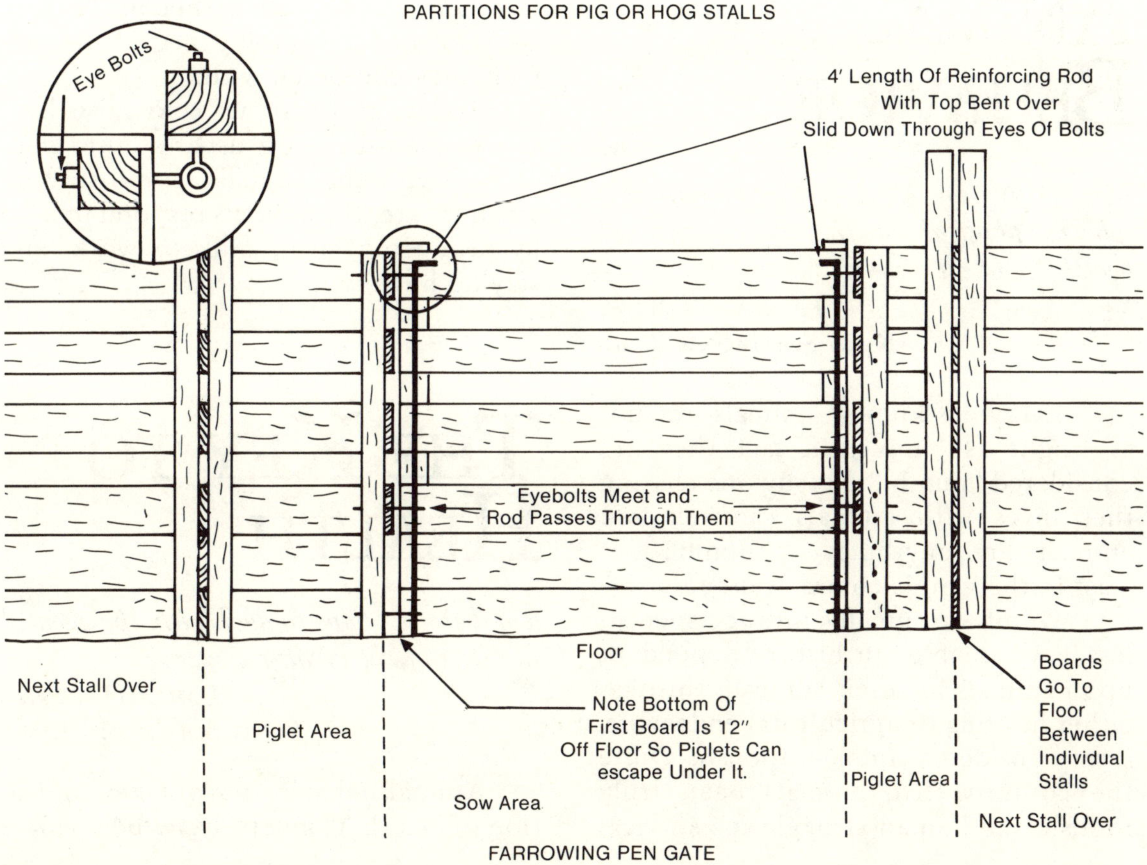

FARROWING PEN GATE

FIGURE 27.

Livestock Management

Animal Behavior

Mental hygiene plays just as important a role as physical in the maintenance of health . . . of all animals.

Konrad Lorenz,
King Solomon's Ring

"Everyone knows" animals are not as bright as people; in fact they are considered "dumb." Yet anyone sharing their lives with animals can tell you animals are anything *but* dumb . . . at least in the word's common usage.

"Dumb" originally meant "speechless." All peoples in history dependent upon animals for their survival, through either hunting or agricultural endeavors, *never* considered animals foolish. Quite the contrary: had animals been truly foolish, the human struggle for survival would have been much easier. Instead, these peoples developed a deep respect for their co-survivalists, becoming very conscious of the intelligence of the animals they competed with.

Today, those of us who raise livestock small-scale are usually quite concerned with the animals' comfort. And when we are, the animals respond in kind by producing meat, milk or work, and staying healthy.

The Key to Health

But here we have to deal with the mental qualities of the creature.

Konrad Lorenz,
King Solomon's Ring

Animal behavior is your key indication of health. Conversely, *mis*behavior is your key to ill health—or at least a problem of the group or individual animal. While limping is a sign of something wrong in an animal, so is a herd animal walking off to be alone; it may be due to imminent parturition, illness, pain, or injury! If an animal constantly makes noise, starts fights, rolls her eyes, or grinds her teeth, this behavior is a sign of the animal's health—and its importance cannot be stressed enough.

Odd behavior is one of the first things your vet will inquire about when a diagnosis is uncertain. It's often the only clue to the nature of an illness. And it can be pretty strange. My vet once told me of a cow who began giving twice the milk she had on previous days. The farmer was delighted until the cow started tearing up concrete and trying to eat it! The diagnosis was acetonemia—and she would have continued suffering from it had her only symptom been the increase in milk production.

Housing affects an animal's behavior. Animals housed in such a way that they are unable to see what is going on in the barn become nervous and in their distress may even create deep ruts around the stall's perimeter. Horses develop classic behavior problems such as weaving and cribbing. Some animals become vicious or deeply depressed from the stress of inadequate housing. And too often these symptoms go unnoticed or ignored. Understanding the animals in your care can make all the difference between good and excellent management of them.

Behavior Extremes

The Queen

We have an ancient Saanen doe named Belle. I have seen her walk through a pen where nine sheep and four other goats were sunning themselves comfortably. Without so much as a glance at them, she would stroll in such a path that each of those animals had to move. Talk about dictatorship!

Years ago, a stray dog bent on violence arrived in the yard just after the goats had managed to get loose. They were essentially defenseless. The goats sighted the dog and immediately bunched with their rears together. At that point, Belle left the group and stalked that dog. She is hornless. True, she could have butted the dog and possibly seriously injured him. But the dog was quick. He continuously tried to break up the group but she remained between.

I have seen Belle repeat such actions in a variety of small and large ways since then and have recognized them for what they are: the expected price of the leader for her privileges. She is fed first and treated with respect.

Fairness and justice are not "natural" concepts; sometimes they happen anyway.

The Underdog

When it comes to the pecking order, the top positions are usually easy to spot. The lower orders can be obscure, but the lowest herd member is easy to detect: she will eat the most, be most inclined to be overweight, and tend towards nervousness.

Because of these things, her diet must be watched. If the herd is very large, the lowest in the pecking order may *not* get enough to eat and may need a little assistance from you. In most cases, she will get too much to eat and gulp food down, occasionally causing digestive upsets. She also tends to be the most vocal in the group.

If the pressure gets too much, she can become truly sick. Splitting the herd into smaller groups will help. (In fact, smaller groups will also quiet down an extremely aggressive queen.)

The Underdog is a natural phenomenon; removing her from the herd will only mean she'll be replaced by another animal. Keep an eye on her and she should be all right.

The Troublemaker

Once in a great while you will acquire an animal who will *not* settle down in her group, refuses to be handled, raises cain when you try to milk or work her, and refuses to cooperate in any way. Cattle people have a term for such animals: bunch quitters.

Be she the Queen Saanen of the goat world, a direct descendant of the great Bast Ra in the cats, Rin Tin Tin's only get, or Sea Bisquit's last living kin, if she creates continual disturbance under the barn roof, you've got a problem.

Approach the situation as calmly as you can. Start with the former owners; did they have the same problem? (Don't be so crass as to ask if that's why they sold her to you!) Did they find anything short of violence that helped to calm the beast?

If this doesn't help, observe her alone and with others of her own kind. She may be having difficulties with new food, or harassment from another animal which starts her off. If she's of a species you are unfamiliar with and you have no others of her kind, something may be wrong with her accommodations. Give her a week to settle down. If that results in no change, consider passing her on.

Rarely—but sometimes—an animal will just not like it at your place. If you cannot supply her with separate quarters or you don't feel she is worth *that* much to you, get rid of her as soon as you can. Eventually, you will anyway, by which time you will have more money and effort sunk into her. It happens to everyone and is no personal reflection on you.

The Cripple

The countryside is dotted with animals missing eyes and legs, or that are blind or deaf or suffering from other ailments and injuries. Generally, it pays to see how they get along with their handicap before making the drastic decision to end it all for them. Allow them plenty of time to adjust—they'll usually do much better than their human counterpart. Of the handicapped animals I've had, the blind cats still hunted, the blind mule and the blind goat continued to work and play in the pasture. In fact, I'm willing to bet that one out of every ten people who raises animals has at least one blind one and doesn't even know it!

Animal Combinations

Even in fairy-tales, animals are portrayed as being a community comparable to that of mankind, as though all species of animals were beings of one and the same family, as humans are.

Konrad Lorenz,
King Solomon's Ring

Single herd animals, or animals that normally function within a group tend to suffer from loneliness. Because of this, if you have a single animal of a herding species, a combination may be necessary. In a pinch, combinations for a few days may solve a brief-lived problem. But as a rule, animals should be separated by species.

Whenever two individuals of different species are going to be housed together, give them a chance to get acquainted in a roomier situation than a stall provides. Penning them together without any previous introduction can spark an immediate hatred, and in territorial defense one may seriously injure or kill the other. In fact, this may happen when a new animal is added to a same-species herd, and should be handled the same way. Give everyone a chance to get acquainted in a looser arrangement, like a pen, for a few afternoons. Then when the aggression is displayed, the new animal has a chance to move away without injury.

Sometimes an animal of a different species won't mix well with a same-species herd. We put a doe in with our sheep once in a pinch. There was no problem outdoors. But at feeding time, the doe took over completely. She intimidated the sheep into remaining on one side of the stall while she ate, and it certainly seemed clear she intended to leave none of it for them! We finally put her in a kidding stall—to the sheeps' relief.

The problem can be reversed, too. If you keep a goat and horse together, you may have a problem separating them when you need to ride or work the horse. If this happens, force the separation and increase its frequency and length as often as you can. Goats don't mind the separation but horses can become very possessive about their stall mates and refuse to leave them. *Do not* house a goat with horns with a horse or any other animal. Sooner or later, there *will* be an accident.

Outdoors, some animals can be combined to each other's benefit. Goats can be run safely with sheep, but watch for potential parasite problems in the goats. Pigs and cows do well, the pigs gleaning food from the cows' manure. Goats and pigs do not do well together; pigs are perfectly capable of arranging for fresh meat—such as newborn kids—and the goats are incapable of defending their young against such formidable foes.

Goats do well with cows, but the cows don't. The goats are too quick for the cows; when a cow lies down in the sun for a nap, the goats use her as a trampoline. I've seen goats climb on horses the same way, although horses don't seem to mind as much as cows do.

All these combinations are possible outdoors with an eye on the individuality of the animals. A horse with a strong herding instinct, for example, may bite or herd the others, causing weight loss for all. And you may have to run a special gating system by the barn to direct each to their own housing. An example of this would be a low opening in one direction for pigs, and a large gate with a twenty-inch board fence across the bottom in another direction for horses (see Figure 28).

In general, run species together outdoors that are dramatically different in size. Goats, sheep, and pigs, because they are all about the same size, can have special problems. With horses and cows it depends on the individual personalities. If you can find a way to keep fowl from contaminating the mammals' water supply, they should do well in almost any combination outdoors. Because of the potential problem with tuberculosis, sheep, beef cattle, horses, and pigs are safer combinations with fowl than dairy animals.

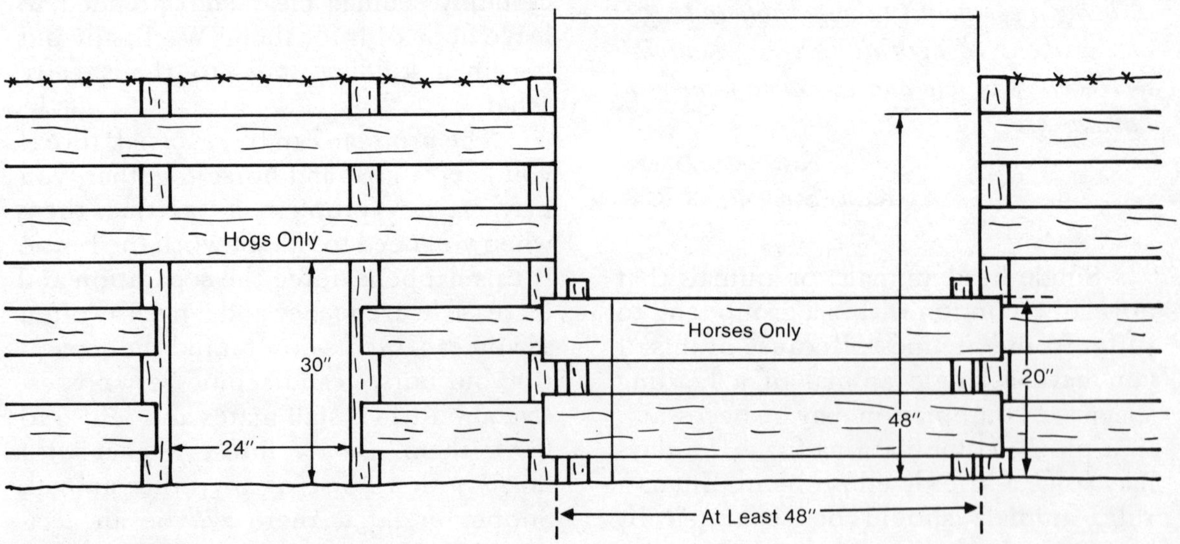

FIGURE 28. HORSE AND HOG GATES FROM COMMON PASTURE

Feeding

Regularity in Feeding

While all the more intelligent species enjoy a change now and then, animals do thrive on routine. This extends into feeding schedules, especially with milking animals, horses, and young of nearly every species.

Regularity in milking schedules can make a pound difference in production daily, and quite a substantial difference in the overall lactation total. The better producers I've owned let their milk down *at milking time*—when I'm ten minutes late, it takes a while to get a good flow going. To change my schedule for a longer period, such as daylight savings time, I do it all at once. In a few days, the stock have adjusted to the new schedule, and five minutes in advance of the new time, they begin reminding me I'm late—just as always.

With milking animals, the best possible schedule is every twelve hours. Many people face the chore of "six and six" with dogged determination. Some folks delight in early morning hours when there is wildlife still around; others dislike the milking schedule and sooner or later, catch themselves oversleeping. This wreaks havoc with the rest of the day's schedule. Some small-scale farmers who enjoy the morning hours in spring dry the milkers off in early November.

Give yourself some consideration in the schedule. If you get up at seven but want to be finished in the evening by five-thirty, so be it. If you can't face chores until you've had coffee or tea first, adjust your schedule; the animals would rather wait the ten extra minutes than be greeted by a grumpy person they've waited all night to see. Dramatic differences in the daily schedule may affect production. But *your* production—and productivity—must also be considered.

At times, circumstances beyond our control alter our schedule. The result may be a drop in milk production for a day or two, and little gain on meat animals. Luckily, things return to normal in no time.

If you regularly feed, milk, and water at odd hours, production and peace suffer. A few species may adjust—perhaps only a few individuals. Generally, animals develop nervous habits such as calling constantly, gnawing wood, thumping the walls, and chewing and fighting with their neighbors. They also produce less than they would on a normal schedule. Whenever possible, maintain regular schedules; they pay off.

•

Efficient systems for animal care and management allow you more time with your animals—on a choice basis. They also mean daily chores can be done more quickly. Over a year you will have saved an amazing amount of time. And an efficient system means you have more time available so that if you find an animal injured, or kidding unexpectedly, you won't have to neglect the other animals. Most important when illness or injury affects a family member, a neighbor can do the chores for you.

Smooth, efficient systems are not a crime.

Salt

Salt aids digestion, appetite, milk production, general health, and good coats of your stock. Use iodized salt in northern climates; the land in those areas lacks iodine. Protected from weather, salt lasts longer and stays cleaner. Supply it on a free basis.

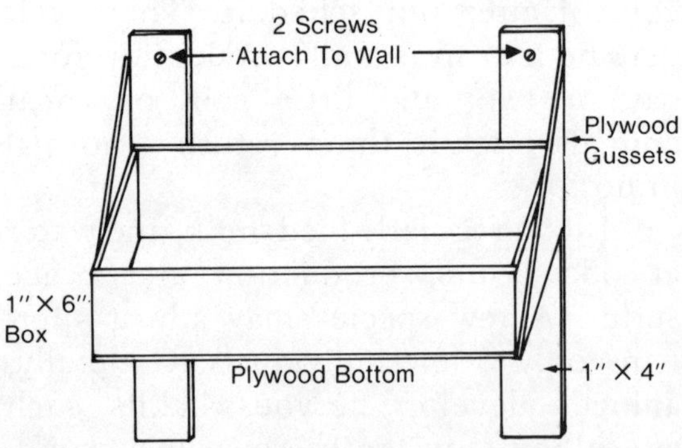

FIGURE 29.

Water

The cheapest quality feed you can provide is fresh water twice daily.

Gene Logsdon

Water is your most valuable feed. And you can produce it yourself, even when you're broke. In its natural form water has minerals that cannot be replaced by feed additives. It's also the one commodity both you and your animals need and can do without for the least amount of time. Water keeps all creatures' systems flushed and clean if they drink enough of it.

Entirely too many people short their stock on water. Dairy farmers who must lug water to heifers located in the farthest part of the barn are often guilty of this. However, if I want an animal to grow well and later to produce milk, I establish in them a firm drinking habit. Occasionally, I've purchased goats that came from places where either insufficient water was supplied or it was often foul. The goats were sold cheaply because of low production. This changed once a better drinking habit was established.

Water should be as clean and fresh as possible. Part of doing chores twice a day should be to change everyone's water. Scrub all watering devices once a week. If you put it off a day or so, water con-

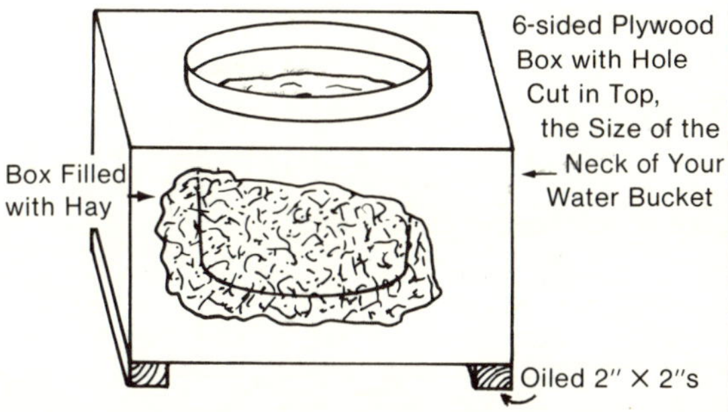

FIGURE 30. FREEZE-RESISTANT STOCK WATERER

sumption is cut but resumes once the cleaning is done.

Animals are most likely to suffer from dehydration in winter—not summer; frozen water is the cause. They *cannot* eat ice and snow and do well. They may *survive*—but that's not the same thing. Kidney failure, digestive problems, and lowered milk production are all probable results of poor water management.

I supply all the water a group of animals can consume in a twelve-hour period. In colder weather, I water them three times or more a day, providing warm water. Only once have I come across an animal who didn't prefer warm water in cold weather, a Californian buck rabbit whose response was to dip his paws in and lick them off. When I supplied him last, the water was sufficiently cool.

If you have some animals that prefer their water with an additive, such as vinegar or molasses, and some animals that don't, you'll have to run two watering supplies, one with and one without the additive.

Water should *not* be in kidding or lambing pens and stalls. Water these animals when they are put in for the night; during the day let them out with the rest of their kind. If you leave a water bucket in with an animal expected to give birth, you may find a drowned newborn in the morning.

Buckets and Pails

Milk pails, disinfectant pails, compost pails for kitchen collection, tanning pails, the inevitable scrap pail for the pigs, and metal ones in which to heat water for the animals in winter—no one has more buckets and pails than the small-stock farmer. For watering, a hose or water hydrant (see Figure 73) is a better system where possible. Much of the year in colder climates, a hose isn't possible, and sometimes animals located off to the side are too far for the hose to reach. Therefore you wind up with buckets to carry water to the animals' buckets.

Plastic pails are probably the most common today. They're terrific for hauling water to animals or for collecting maple syrup. But they crack under even the mildest strain in cold weather. At best, they are difficult to disinfect, and they wear out quicker than any other sort. They absorb odor, so are out of the question as milk pails. *Beware of plastic pails that originally contained poisons.*

Stainless steel seamless pails are the best for milking by far. But oh! the cost! If you can't bring yourself to pay for one, many feed mills sell two-gallon aluminum pails for just such a purpose. Seamless are easier to clean. To keep bacteria down, cut the wooden grip off if the pail comes with one.

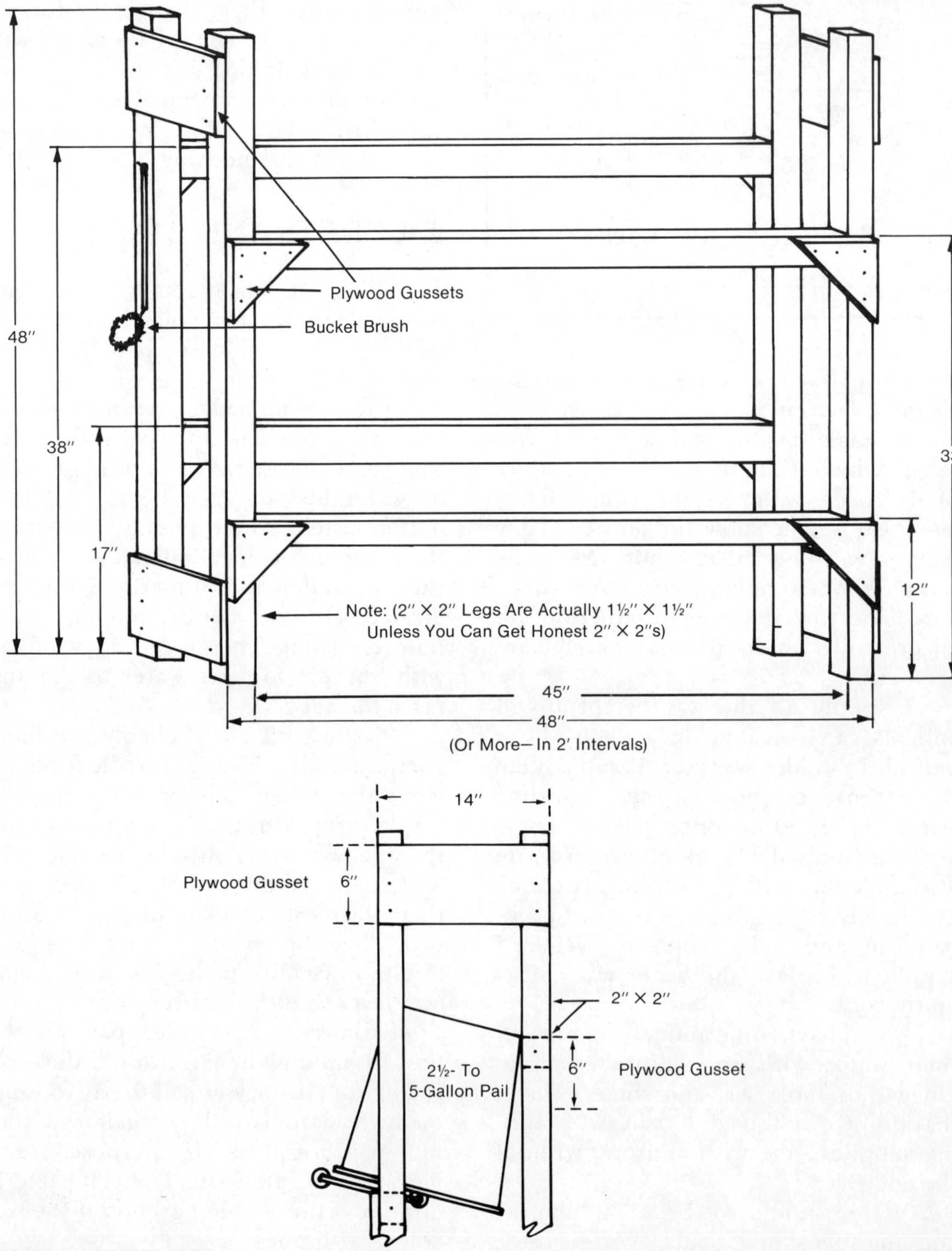

FIGURE 31. BUCKET RACK

The aluminum pail has one drawback: lactic acid reacts with aluminum; if milk is left standing in such a pail, it may change quality. But you should strain your milk into a glass container as soon as possible anyway from *any* kind of pail to keep it clean.

Metal pails (of the normal kind) have advantages for general work: they can be heated and take a beating from livestock. But the seams on the standard metal pail are impossible to clean and the pails disintegrate from day of purchase. Animals may injure themselves on them, banging a foreleg or scraping an udder.

As watering devices, rubber buckets are best. They last just short of forever in heavy daily use, and they scrub up well with a brush. They make safe grain buckets for horses and pigs, and animals can't get hurt rough-housing with an empty pail. You can bash ice out of them winter after winter and they spring right back. The initial cost is higher than for any other pail. It seems worth it somehow.

Bottle/Pan Feeding of Kids and Calves

Bottle feeding is the most natural for kids and calves. Nursing from their dams, their heads are tilted up to aid digestion; follow nature by feeding them in this position. We bottle feed all kids and calves to get them off to a better start, then switch to pan feeding. Advocates of bottle feeding claim they are less likely to suffer bloat this way.

Sanitation is a must. Infant animals ingesting foreign bodies such as hay or manure via their milk can become quite ill. More frequent feedings also decrease the potential for bloat; with either method—pan or bottle—the more often they are fed, the better.

To decide whether to bottle or pan feed, let's take a theoretical group of kids, say three, and feed them a minimum of three times a day. That totals out to three bottles, three nipples and a measuring cup and funnel, plus a pan to heat the bottles in for each feeding. It isn't necessary to wash the pan the bottles are heated in each time, which leaves eight items washed three times a day for a total of twenty-four per day. At the end of the week, you have washed one-hundred and sixty-eight items for just three kids or calves!

We feed our kids and calves four to five times a day for three to five days—the variation results from how the animal is doing, the weather, etc. Then we switch to pan or pail feeding. We begin with a single pan per three kids. For comparison's sake, feeding the same three kids three times a day in a pan of measured milk means two items to wash three times a day, for a total of six. Compare that to twenty-four with bottle feeding. And the week's total for pan feeding is forty-two, quite enough in itself! Many pails have measurements imprinted right into the metal or plastic, so no measuring cup is needed. Use strictly metal pail and pans for feeding young because you can heat them directly on the stove. For kids, any cleanable flat pan will do.

The stall doors on kid and calf pens are six inches above floor level so a pan can be slid under the door. Any draft is prevented by folding a feed sack in four lengthwise. Then nail the sack to the bottom of the door with a strip of wood

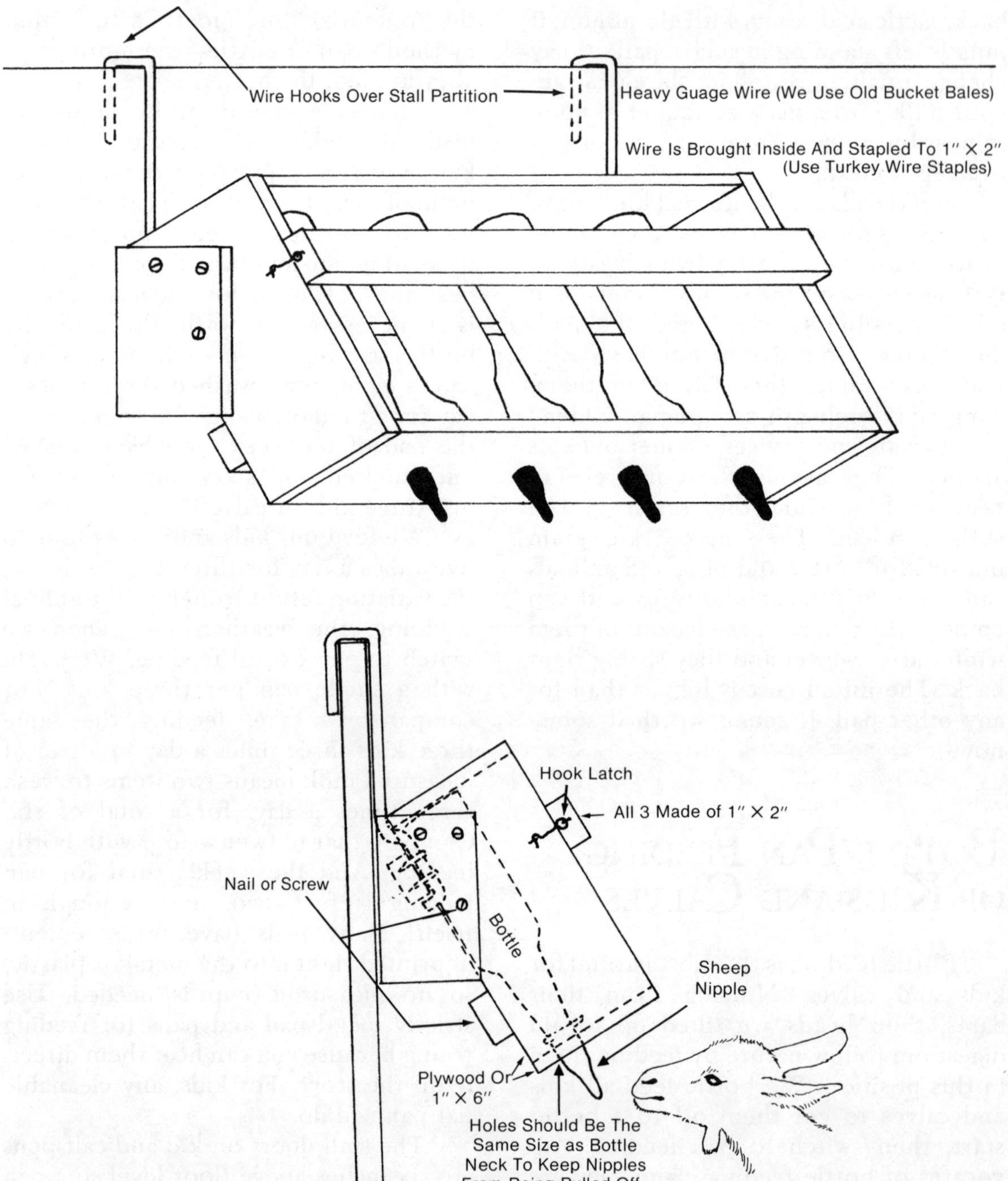

FIGURE 32. BOTTLE HOLDER

over it to prevent the nails from pulling out. It should reach the floor when the door is closed (see Figure 60). Lift the sack to slip pan under, and leave the pan edge directly under the door to prevent tipping. It's easy to change sacks when dirty. We use them on all stall doors in winter to cut drafts in the barn.

The bottle racks I've seen show the nipples free on the kids' or calves' side. Young animals are strong, and the kids often pull the nipples off and keep sucking on them while the milk spills all over. In fact, before I could prevent it, one kid had nearly sucked it down her throat. The bottle rack in Figure 32 is designed so that the kids *cannot* pull the nipples off.

We keep kids on milk longer than average, three to four months for doelings. We feed them milk three times a day until two months of age and, thereafter, twice daily. We add extra water to encourage good drinking capacity and use hot water as the heat source of the milk from two months on. We provide fresh water, hay, and grain from one week of age on. This arrangement is best for us but may not be for you. Make a schedule to suit your own routine.

Free Feeding

Cats and Dogs

Cats and dogs will eat *less* on a free-feed basis than on a rationed diet. Dry food works best and on a free-feed basis, where the food is set out so that the animals have free access to it. Consumption will go up slightly in foul weather, because idle animals snack as much as idle people do. Supplement their regular diet with leavings from butchering, but pay attention to the pest cycle (see p. 86).

Cats need a higher protein and fat ration than dogs do. Cats also need some amino complexes not found in dog (or many human) foods. Such is the case with *taurine*, necessary for retinal function. (Cats go blind without it.) Dietary additives for these two species such as table scraps, whey from cheese-making, and eggs (cooked; raw eggs destroy the B vitamin complex) are fine. In fact, a varied diet ensures that your animal receives a full spectrum of vitamins and minerals.

As a rule, cheaper dry foods contain cheaper ingredients, but most are adequate. Avoid softer brands of dry food. They cake up on animal's teeth at the gumline and cause problems.

Provide a bucket of water, changed daily, for both cats and dogs. It's amazing how much water cats will drink, given half a chance.

Occasionally, we've had some problems with flies in free-feeding arrangements, so check pans regularly. Clean and wash pans with a disinfectant, as with all other animal buckets and pans.

Also watch for raiders. A duck bullied one of our more tolerant dogs and ate all her food. We even surprised an opossum raiding the cats' food. Moving the pan to a new location will usually solve the problem. Catching and taking the opossum for a drive didn't hurt either.

Livestock

Free-feeding livestock has its problems. Grain is fed in controlled amounts for several reasons. Given free access, many animals (such as the horse, but excluding the pig) will eat until the results are serious, if not fatal. I'm referring to mature animals normally fed in regular amounts, not young animals on creep feeding.

Animals like grain because it has what they need. Animals seek needed nutrients, and grain supplies many of them. But grain in its present form is a relative newcomer to the feeding of livestock.

Many mixed feeds contain a great deal of molasses. Feeds like oats have a natural, tasty sweetness to them. If grains are fed whole, large amounts will pass through the system intact and little or no nutritional value will be had by the animal consuming them. Grains must be cracked, ground, rolled, or steamed to break or soften the hard outer shell.

When grain is fed to animals in small amounts, perhaps because they know they can't have it as often as they'd like, they gobble it down when the opportunity presents itself. This also results because of the competitiveness within the herd. Some animals will eat more than they really need, knowing that another animal will get it if they don't clean it up.

People who have tried free-feeding have found that most livestock overeat at first (dangerous enough!) and then taper off to little if any interest in grain. The initial period is dangerous because of bloat, founder, enterotoxemia, and bowel impaction. Until more research is done, it's wise to stick with rationed grain for mature animals.

Hay is another matter entirely. In their natural state, sheep, goats, cattle, and horses are forced to move around a great deal in the process of eating, thereby getting lots of exercise. In domestication, however, they often stand in one place for a long time. Nervous energy, caused by nearby competition, means some animals eat more than they actually want. In any group of more than three animals of the same species, the lowest member will overeat because she feels constantly threatened. The herd leader may overeat because she fears another animal will gain strength and take her position. Sometimes this overeating results in increased production and can be used to your advantage. Sometimes not.

"Feed all they will clean up at one time" was something we never tried on animals other than horses until recently. We now feed hay to our goats three times a day in the coldest weather when pasture isn't available because of snow cover. The rest of the time, we feed all they will clean up twice a day. Not only do they consume about the same amount of hay (keeping up production) but they also get more exercise by searching the pasture for succulents and "something to gnaw on" all day. The exercise improves their general health, and they're all friskier. They don't hang around the barn anymore and are in prime condition. Best of all, there are no hay piles all over the floor.

Grains in their present forms are

relatively new to the management scene. Some feeds you can raise yourself include the following.

MANGELS are a large, bulky beet grown specifically for livestock. In fact, they often are included in the bulk of animal feeds, and are high in minerals.

CABBAGE is very nutritious for both animals and people, but must be fed with care as it tends to cause bloat.

PEAS AND BEAN VINES were often used as a summer feed for horses and other livestock. They increase milk production, as do all legumes. Because they are very high in nitrogen, watch for scours.

PUMPKINS AND SQUASH increase the butterfat in milking animals. They are a good solid feed which can be used as the bulk of a pig feed. Poultry love them cooked.

CARROTS are well-liked and nutritious. They also improve production and butterfat in milkers.

COMFREY is an ideal animal feed. It's the only ground plant known to carry the vitamin B complex.

JERUSALEM ARTICHOKES AND POTATOES can be fed to animals as well as your family. They are often used cooked as the bulk of feed for pigs and hogs.

SUNFLOWERS can be fed to all animals and are preferred by fowl. They are making a comeback these days.

CORN can be harvested by hand and stored easily for all animals. Caution with horses: corn is a heat-producing grain and can cause blisters and heat prostration in equines in summer. Corn, stalks and all, can be shocked and fed over winter.

All of the above feeds store easily over the winter and add variety to your livestock's diet.

Introduce new feeds, whether formulated or garden produce, to your stock slowly. Watch for reactions such as bloat or scours.

Other garden produce can be fed, with care, to livestock all through the gardening season.

PREFERENCES

FOWL prefer anything green and leafy but will often eat cucumbers, tomatoes, and melons as well. They love sprouts of any kind.

PIGS' natural feeds are root crops, but they enjoy many ground crops as well, such as squash.

GOATS AND SHEEP like a little of everything. Feed surplus food to goats in a rack: they won't eat anything that falls to the ground.

Many weeds pulled from the garden are good for the stock, too. For example, lamb's quarters' seed tops contain 26 percent protein. But be sure of what you're feeding them; some weeds are toxic to some species.

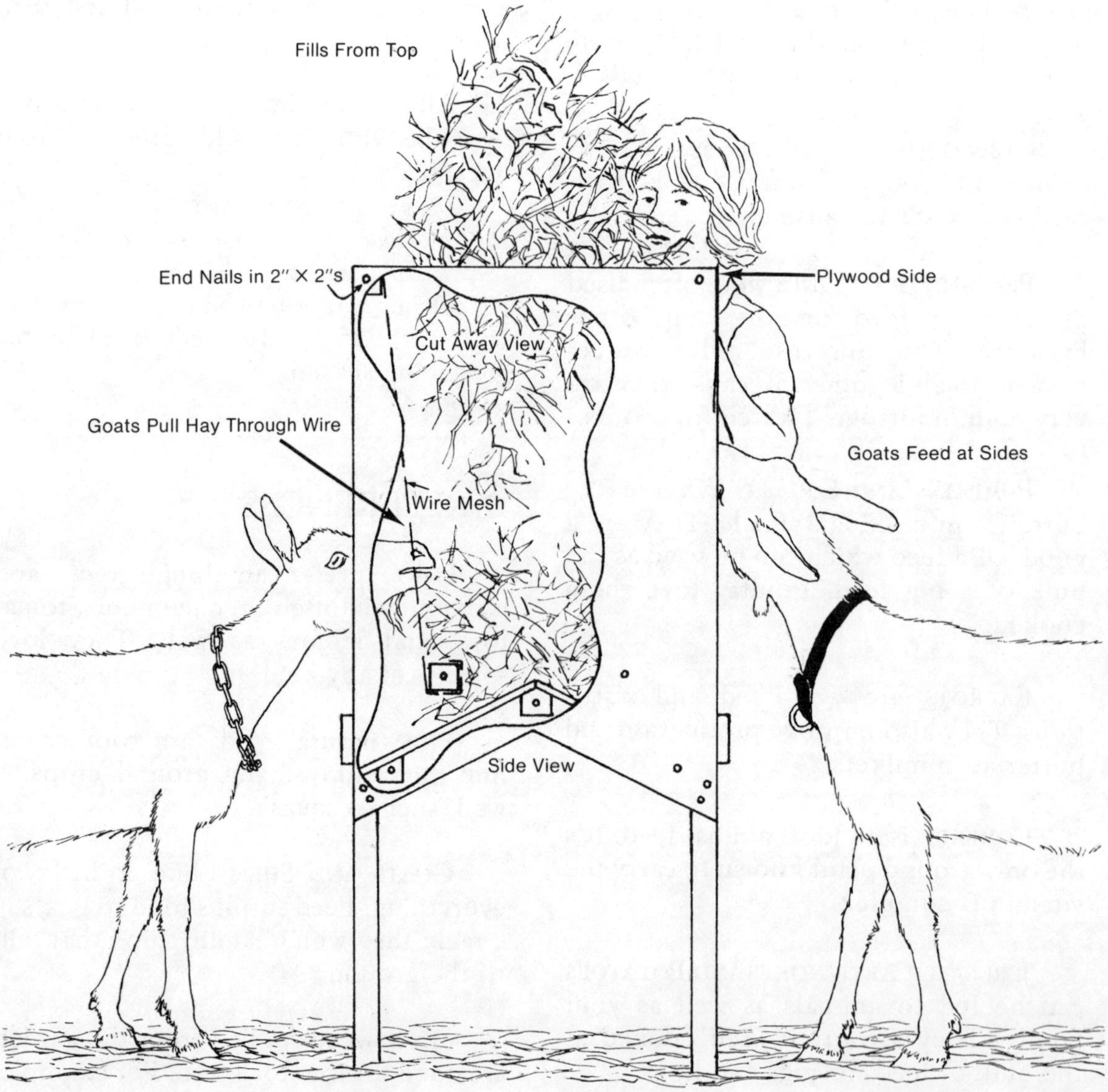

FIGURE 33. GARDEN WASTE AND HAY FEEDER

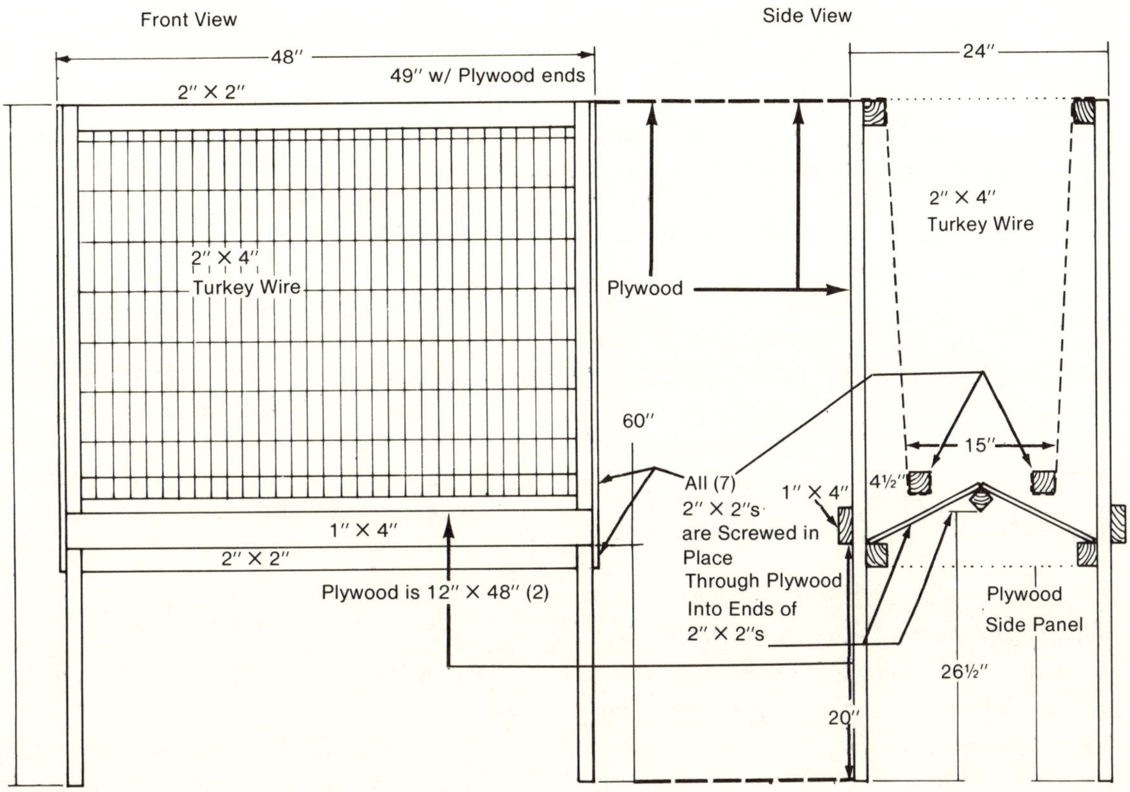

FIGURE 34. GARDEN WASTE AND HAY FEEDER (detail)

Free Housing

I would feel more optimistic about a bright future for man if he spent less time proving that he can outwit Nature and more time tasting her sweetness and respecting her seniority.

E. B. White,
Points of My Compass

Free housing allows animals to go outdoors or in at will during the day. This arrangement increases the chances of good health because it permits animals to change their environments at will. Social pressures are reduced when an animal can escape through an open door. And the whole group can go in when they feel lazy, or out when they feel the need to exercise—instead of tearing the indoors apart.

Free housing decreases the human work load. Driving or leading a group of animals back and forth to pasture every day is time-consuming and frustrating.

When I first had goats, I learned they became frantic when a storm threatened. As I dashed out to run them to the barn, one goat would always rush off elsewhere. Leaving them out was no answer either; one doeling became so frightened, she made it over the fence and didn't return for two days.

Free housing also decreases dependence on you—always a good thing—and gives the animals a choice of comforts.

Confinement

Only under the unnatural conditions of close confinement does it become apparent that [some species have] no inhibitions which prevent [them] from injuring or even torturing [their] own kind.

Konrad Lorenz,
King Solomon's Ring

Most domestic animals have no inhibitions against injuring their own kind. The chicken being pecked by her neighbor simply moves away; in confinement, she cannot and the result is cannibalism. In the wild, the hornless goat flees her horned attacker; in confinement, she may not be able to. Death will result in a loss for you instead of just food for the carrion eaters.

To guard against the problems of confinement, take the following precautions:

1. Plan sufficient indoor space for each animal and species.
2. Castrate all male animals not used for breeding purposes.
3. House separately breeding males except during breeding season. As an example, a doe may not be seriously injured when the buck slams her against the wall, but her milk production may decrease. The dairy goat is awkward with her big udder and cannot escape injury as easily.
4. Allow as much room outdoors as you can afford for each species.
5. Disbud kids and calves. The stress when done, is brief compared to a lifetime of getting caught in fences and battling with other horned animals. Floor space must be three times as large for horned animals also, which increases the cost.

Escape

Sometimes, it's necessary to keep animals confined for longer-than-normal periods. Foul weather, emergencies, or immediate availability (the vet is coming but you don't know exactly when) are usually the circumstances. To entertain themselves, penned animals often find something to do such as systematically pummeling the lowest member of the group or escaping.

The word "escape" implies captivity, which is only partly true for livestock. The average domestic animal looks to you for food, shelter, and medical aid when she's in trouble—she's done so for millennia. Even when she breaks out of confinement and spends the day dismembering your hard work, she'll be around at chore time trying to mooch her grain ration. Animals are prisoners only in the sense that they depend on *you*.

Teaching Animals to Escape

By nature, pigs and hogs burrow to escape. However, through clever fencing and short-sightedness, we once taught a very large hog to hop fences like a steeple-chaser—and you might heed how it was done so you can avoid the same miracle.

We built a justifiably stout pen of 2 × 6s about thirty inches high, and loaded our two new feeders inside. In two days, we discovered one of the hogs had a knack for jumping. Instead of really raising the pen sides, we raised them six inches with another bar—the same method, it so happens, that is employed by trainers of jumping horses. And since the hog had all day to practice, he learned pretty quickly to jump that top bar. Quite a few hours of fencing could have been accomplished in the time we spent pig-chasing. When the pen was finally high enough so he could no longer jump it, he had learned to climb it in the corners. Before it was over, we had a five-foot-high pig pen with a strand of barbed wire along the top—and funny looks from pig-raising friends who saw our fortress.

Only after the pig reached two-hundred pounds did he accept defeat, but we were too weary from pig-chasing to enjoy the victory.

If you've got an escape-type, stop what you're doing, figure out what will stop them for *sure*, and do it right away. Do not teach them better tricks. For the sake of the human reputation, get it right the first time.

Boredom

Now which are the animals really to be pitied in captivity? . . . Those clever and highly developed beings whose lively mentality and urge for activity can find no outlet [in close confinement].

KONRAD LORENZ,
KING SOLOMON'S RING

Adequate space usually relieves an animal's boredom. Animals will entertain themselves if given some companionship, enough food and water, and as much space outdoors as possible. Fresh air, sunshine, and room to exercise keep them quite happy.

Boredom results from too small a space. Animals will make every effort to escape if they are bored; they will become unusually destructive, aggressive, noisy, and depressed. Boredom can lead to a poor appetite and begin a cycle of either bad manners or even a health problem.

When they can't be out of doors all the time because of foul weather, or there is some reason to confine them, or you just don't have the room or finances to provide them with more space, there are things which will help.

For pigs, cast an old tire into their pen. They'll root it for a while, flip it into the air, then run around it gleefully. Pigs are very intelligent, and do better raised in pairs for company. But a tire will help a lone pig keep her mental health. And that means better production in the long run.

We've had several geese fall in love with hanging tires. Goatlings will swing on them—but that's outdoors. Indoors, an oatmeal container suspended on bailing twine just high enough so they can reach it with their nose tip will keep many animals entertained for hours. It smells tantalizing—but when they bump it, it moves away.

Two tires connected by a short rope and tossed over a horse stall partition (with a tire on either side) will keep horses preoccupied for hours in a harmless way—instead of swaying, wood chewing, cribbing, and a host of other things horses come up with to combat boredom.

If close confinement is necessary for long periods, provide more than one outlet for their attentions. They need variety. And again, provide as much area as possible.

Stocking Up

There's another physical difficulty that occurs in livestock animals when they are cooped up without exercise for even moderately short periods. Horse people call it "stocking up." Smaller animals suffer from it less than the larger animals, because small animals can get at least marginal exercise in a moderately-sized stall.

Fluid builds up in the pasterns of a quadruped when they get little or insufficient exercise. The cure is simply reversing the cause.

Give all your stock the opportunity for sufficient exercise.

Bedding

Bedding is absorbent material used to keep the animals clean. It should be undesirable to eat; when bedding is of a material that an animal might eat, your chances of parasite problems are greater. If you make compost, bedding of a carbonaceous material is of more value in the heap. Bedding that is somewhat "crushable," is softer to lie on and reduces skin problems. A comparison of the more common types of bedding might be helpful.

Straw is the best. It contains a light oil that keeps an animal's coat shiny and clean. Once you get the hang of separating out the soiled straw with a pitchfork, it's an easy daily chore to keep a stall in good condition. Straw is undesirable to eat, although most animals will taste a few straws when the stall is freshly bedded. It's carbonaceous for the compost and easy to handle by the bale.

Hay as bedding is common. In a pinch we have used waste hay—floor sweepings already rejected by the stock. Many people use moldy hay, which has serious consequences if the stock eats it. Non-legumes are carbonaceous for the compost but hay is also full of seed, and makes a lot of weeds. It's also a potential food.

Wood Shavings are good. They must

be purchased, which means you must depend on a company's cooperation and reliability. Be careful of what kind of wood it is; some woods, like cherry, can be toxic.

WOOD CHIPS are *not* good. They are not small enough to be comfortable; they tend to mold and mildew, creating dampness instead of absorbency. They also may contain toxic woods, such as cherry, but they are good for the compost.

SAWDUST is also good and its production is as dependable as lumber. Your major competition is the toilet paper industry. In your compost, it increases the acidity of both the compost and the garden. Adding lime will help. Sawdust cleans easily and wet spots can be picked out daily.

SUGAR CANE, sold by the bale, is expensive. It's carbonaceous but not very soft or absorbent.

DRY PINE NEEDLES are difficult to collect in sufficient amounts and are hard to store for later use. They seem to disappear in a stall in short order, although the animals don't seem to be eating them. I assume they disintegrate. They are suitable in the compost, but extra lime may be needed. They're free, if you've got the time to collect them.

SAND as a bedding base holds urine instead of draining it. And it smells something ferocious at the end of a week.

Cleaning Stalls

Goats, sheep, beef cattle, and rabbits (if in a stall arrangement) should have their stalls cleaned as needed, with a major cleaning in spring and fall. Remove bedding completely and apply lime before new bedding is spread. Top-dress stalls, as needed.

Goats will keep themselves as clean as they can, given half a chance.

Sheep need frequent top-dressing if you expect the wool to stay clean in a crowded stall.

Beef cattle, if allowed outdoors, can still build manure at a fantastic rate indoors. It may reach the ceiling before spring cleaning.

Nominally clean horses at least daily. Top-dressing with fresh bedding is usually necessary every forty-eight hours. We thoroughly cleaned horse stalls every week, but you can get by for longer if you top-dress.

Dairy cows need a great deal of special attention when kept in a stall. Their manure is soupy, and you expect them to be clean enough to milk. Although you see cows in large numbers on concrete, the well-bedded stall is much better for them. Don't skimp on bedding; if she falls, she'll break something. Knees, brisket, and elbows suffer in a thinly-bedded stall. We were happiest with sawdust for the dairy cow because the "pies" and wet spots picked up so easily every day.

Keep hogs on bare concrete and bed

a sleeping area for them. The hog's head is so low that the moisture in any bedding keeps them constantly breathing damp air. Hogs also wet their bedding by vigorous splashing at meals. Most hogs will keep a "bathroom" area in their pens and never soil in the rest of the stall.

As part of your cleaning program remove all manure piles from livestock quarters immediately. Parasites get right back into your animals via such piles. Bacteria love them. Goats like to climb on them, which makes them the most parasitized animal in the barnyard. Piles also harbor hatching flies.

Long ago, farmers kept manure piles just outside the barn as a heat source. As a rule, their presumed benefit doesn't outweigh their problems.

THE MEDICINE BOX

Animal medicines can be stored in the barn or in the house. House storage is inconvenient. The extra trip may mean a small cut goes untreated. If it becomes infected, there's a double nuisance; but house storage insures the animals can't get into the medicine supply.

Barn storage is often inefficient because of lack of planning. Animals can easily break into a thrown-together arrangement, and extra attention may be necessary to keep things clean.

Locate medicines where animals can't get into them. Any area already designated as "Human Use Only" is best, such as a gated area—perhaps near the milk stand. Then don't forget to shut the gate.

I have a sturdy, shallow and wide box on the wall near the milking stand. I nailed a 2 × 4 to the joist above, just long enough to be within a few inches of the top of the box. To the end of this 2 × 4, I attached a catch, matching it with a magnet on the lift-up door on the box. I can open the box without hands by bumping the magnet with an arm and breaking the connection, then raise it with an elbow to the ceiling attachment. I take what I want and elbow it shut again. It's ideal when you are doing something that demands clean hands, especially when no help is available.

The size and contents of the medical box depends on how much of your own medical work you do. Two points of advice: make it a little bigger than you think you'll need to allow yourself some room to grow. Also, make it shallow and wide instead of deep and narrow. Otherwise, you'll find yourself always rummaging for things in the back. Shallow boxes keep everything up front where you can see them easily.

We keep a roll of paper towels alongside it on the wall. On the shelves below we store electric shears, brushes, hoof picks, etc. In the box, there's a shelf strictly for large quantities of delousers, shampoos, and wormers. The next shelf is exclusively for pathogenic medicines.

On the last shelf, the instruments and medicines have increased over the years. A few really bad situations where,

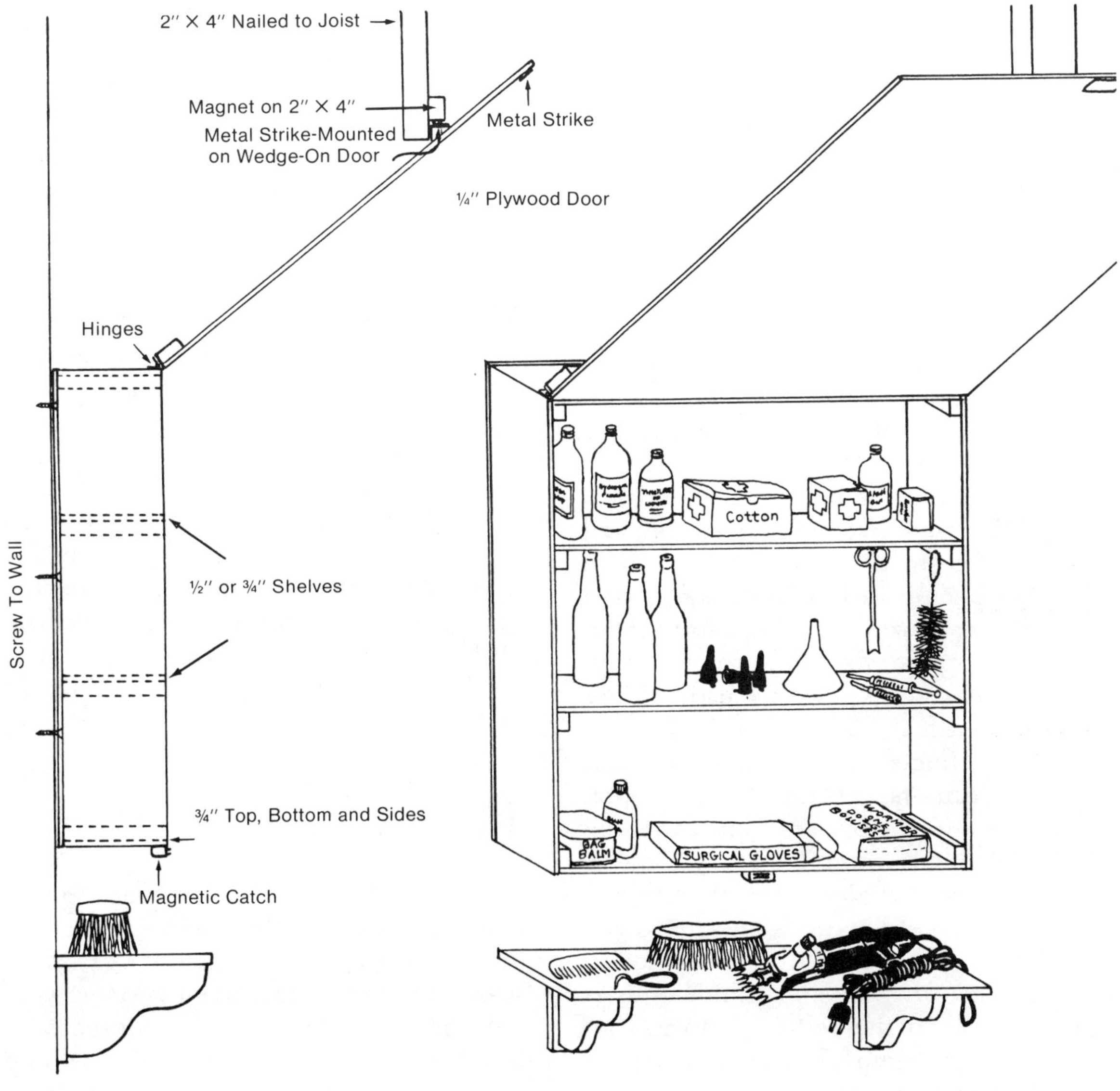

FIGURE 35. MEDICINE BOX

for one reason or another, a vet could not be reached and the life of an animal depended on immediate action has made the contents of this shelf dramatically multiply. Yours may be much less complicated. Mine contains, among other things, iodine, sterile gloves, stethoscope, scalpel, sewing and hypodermic needles, and syringes. I don't pretend to be highly skilled with these instruments, but when a veterinarian is unavailable and the situation is going to end in a fatality if no action is taken, I'll try. Such situations are wonderfully rare.

It's also a good idea to keep your vet's phone number and *two* alternates listed near the phone as well as on the box.

The Value of Touch

The value of touch in animal care is often underrated or ignored. Casual touching is important to the animal as a confidence-building communication, and one that livestock are more familiar with than people are. It has another function as well: heavy winter coats, busy schedules, and poor visibility don't always make it easy to know the little things going on with an animal's health.

Your fingertips will feel the less obvious tumors, swellings, cuts, abrasions, heat from bruises, and other injuries. Animals cannot do anything to relieve these problems for themselves. And they can't tell you—so it's up to you to "watch" for them with your hands.

The majority of animals hide injuries in response to the self-preservation instinct. If an enemy knows you are injured, it can mean defeat. Long ago, I became aware of this when caring for a Canada Goose which had wandered into the yard with a broken wing. We put the wing in a splint, and lightened the splint regularly as the healing process continued. Eventually she was able to continue her flight south of her own accord.

While writing an article about the process of healing and releasing a wild creature, I attempted to photograph the goose and her splinted wing. I spent two long, arduous afternoons attempting to get the photo, but she continually presented me with her good side. I began to realize the same thing happened when feeding her. As soon as the wing was without bandages and functioning up to par, I could have all the shots I wanted. I have since noticed this behavior in domestic animals. It makes sense when you think about it, and is something to watch for.

Examine your livestock daily. Check them by touch weekly; some things won't show up unless you do. I've located unexpected tumors, puncture wounds in hoofs where the heat of the hoof was the only clue, and large cuts that were invisible two feet away. Permanent damage can be prevented by locating these things early.

Blood

Watch for blood in the tiniest quantities in the barn. Many animals clean it up. Even a spot is sufficient reason to search an animal thoroughly—even *after* you think you've found the injured one. Because animals make an effort to hide injury, you may find you've been handling the animal for days without a clue.

As an example, our dog was leaving bloody footprints. Checking the foot produced no results. I washed the foot off and found it to be completely free of injury. It turned out to be another foot which she had cleaned up carefully after tearing a nail completely off; she'd stepped in the blood with her good foot.

Reducing Risk of Disease

Dirty, wet, cold, crowded animals are stressed animals and they get you in health trouble every time.

GENE LOGSDON

Almost all diseases are caused by the same conditions and can be made either less likely or not even possible by good management.

1. Too many of one species creates its own problems. Sanitation and vaccination are the keys here, as well as stress-free housing.

2. Crowding causes stress which may manifest itself as a social order problem. Cramped housing cannot keep animals healthy.

3. Dampness is a major cause of illness, and if coupled with drafts, the whole group will be unhealthy. Too warm and/or crowding increases the potential for sickness.

4. Introducing new animals into a herd or group without an initial ten to twenty-one days of isolation is just inviting trouble.

5. Parasite infestations not promptly attended to within the housing or in the pasture, weaken animals and make them more susceptible to anything that comes along.

6. Animals confined for too long without exercise are susceptible to health problems as well as petty wars staged by bored animals.

7. Stray dogs can bring in diseases that you wouldn't normally have.

8. Mixing young and old animals of any species can create specialized health problems. At least keep them separated until the young are past the most critical age—six months as an average.

9. Keep bucks, rams, boars, and any other male animals separate from the herd if they are used to service outside animals. (In your breeding fees consider the disease potential.)

10. Thoroughly douse the navels of all newborn animals with iodine as you would an open wound.

When possible, buy only healthy, tested animals.

Diseases Communicable to Humans from Animals—and Vice Versa

RABIES: Dogs and cats can transmit this disease to you. All mammals have this potential, although somewhat lessened in livestock. Vaccinate against it as recommended by your veterinarian. It's cheap and easy to do. The disease is controlled now because people took the time to have animals vaccinated. Let's keep it that way.

RINGWORM: Ringworm is a fungal disease. Any mammal can get it and pass it on. I've witnessed one incident where it was clearly communicated *to* animals from a *human*.

SOREMOUTH: Soremouth in sheep can be transmitted to humans. If there have been no reported cases in your area, your vet may see no need to vaccinate against it. But *be sure* to tell your vet if you bring animals in from *outside* the area.

BRUCELLOSIS: This disease is making a comeback. It was nearly eradicated in many states by a simple vaccine; people stopped worrying about it, stopped vaccinating against it, and so the comeback. It affects cattle, swine, goats, and dogs, and can affect horses and sheep. In humans, it's called undulant fever and is most commonly transmitted by drinking unpasteurized milk. In the last forty years there are no recorded cases of it in goats.

Testing milk animals for this disease once a year is adequate protection. Tests for brucellosis are normally given at the same time as tests for tuberculosis. State funds may pay for the tests if you sell the milk. Check with your vet to see whether you qualify.

TUBERCULOSIS: Cattle are usually the carriers of tuberculosis but it can be transmitted to other animals. If fowl are carrying the avian form of the disease—which is *not* transmittable to humans—the cows or goats may test positive on a yearly examination. If they do, *it's the law* that the animal be slaughtered, even though they are carriers but *not infected* animals. Since fowl can usually be replaced at far less cost than a cow or goat (or herds of each), it seems wiser to just keep them separate at all times.

Ask your vet if you should vaccinate against transmissable gastroenteritis (TGE) if you have only a few pigs. And while you've got the vet on the phone, ask about hog cholera vaccine as well. If you live in an area where they aren't a problem, you may find the expense unjustified—although it's minimal compared to the cost of treating a sick animal.

DISEASES TRANSMITTABLE FROM SPECIES TO SPECIES

MANGE: Mange is transmittable between species. Goats frequently contract it from pigs.

DISTEMPER: There is no medical resemblance between the dog and cat varieties of distemper. (Cat distemper is really panleukopenia.) Both are highly contagious, serious illnesses. A cheap vaccine makes all the difference. While you've got the dog at the vet's, get her vaccinated against *leptospirosis* and *hepatitis* as well—both are communicable and serious. And if you have a horse, ask your vet if the dog needs a tetanus shot. You've already paid for the visit and the vaccinations are cheap.

HEMORRAEGIC SEPTICEMIA. This disease may spread from one species to another, as well as from one animal to another in a given group. Commonly known as *shipping fever complex*, an animal does *not* have to be shipped to contract it. The initial cause is stress of any kind. Since shipping is a major stress to animals, its common name is understandable.

(Some diseases are an integral part of the soil in given locations. With this in mind, ask the former owners of your homestead if they had any particular problems. If Johne's disease, blackleg, or Bovine Virus Diarrhea are mentioned, ask a vet's advice with a thought to possible prevention.)

Reduce stress in your animals wherever possible. When it's necessary to cause them stress—such as a visit from the vet—be certain the visit is really necessary.

Tension can also be reduced by varying your pace. Two examples will help clarify: Our vet moves quickly and efficiently with animals resisting treatment. When it's over, he stops to fuss with the animal a bit, leaving them with that friendly contact as the last memory. They are less alarmed when they next see him.

As a second example, if we get a lot of resistance from goats or sheep not used to having their hooves trimmed, we slow the pace between hooves. In between we give them a little bit of grain and scratch their heads. While trimming, though, we speed up and do that animal more often instead of doing a perfect job the first time. It keeps the feet adequately trimmed and at the same time the animal is getting used to being handled frequently. It's a little longer at first but a lot quicker in the long run.

If you must handle an animal through a stressful situation, *calm yourself first*. Your nervousness is easily sensed by the animal, which increases her alarm. From the animal's point of view, you are big, strong, and in charge; therefore if something makes you nervous, it must really be bad.

Don't let very vocal animals or frightened animals stampede you into an action you'll regret. Your intelligence and forethought diminish in direct proportion to the height of your nervousness. Take the time to be calm. It can mean the difference between success and failure.

General Pointers

Avoid allowing strangers with "colds" into the barn; especially don't let them handle newborns. With all modern medicine has accomplished, there is no cure for the common cold, and young or infant animals do pick them up from people. Courtesy and firm insistence usually work; where they don't, toss the offending party out on his or her ear.

It's not a good idea to go straight from a livestock auction into your own barn. Diseases can be carried on your hands, clothes, and shoes into your barn. Vets often scrub their boots upon entering and leaving each barn.

Horses create a susceptibility in themselves and other animals to tetanus. If horses have been kept in your barn in the past, ask your vet which species should be innoculated. Dogs, goats, and people are more susceptible, but your vet may feel other species need it as well.

As long as fowl are kept out of the barn, manure piles are kept away from all animals, things are scrubbed regularly, and species are separated by stalls, your problems will be small.

If something highly contagious or else not satisfactorily diagnosed is going around in your barn, take every precaution you can think of. Change your clothes before entering or hang a pair of bib overalls just inside the door. If a problem continues and the source is still unknown, keep a footbath inside the door and rinse your boots on the way in and out. A cheap cat-litter pan works well, or any shallow pan wide enough for a boot to be rinsed in.

Take heart—these situations are almost unheard of.

Birth Control

Probably the most important item in the management of cats and dogs is birth control. The number of dogs and cats in the United States is *phenomenal*. Because of this, they are a "free" commodity and those less scrupulous feel free to exploit them or let them needlessly suffer. The way to eventually stop this callousness is to exercise vigorous birth control.

General Management

Cleaning

Parasites in the Barn

For farmers with mixed-species housing, external parasites can be a serious problem. Because disease is transferred easily from one animal to another by flies and mosquitoes, every effort should be made to control and exclude them from your barn. Screened windows are beneficial where possible—even with open doors in the barn. Flies enter through top openings in buildings; shorter doors for shorter animals will help.

Some animals attract flies more than others. Horses have that problem more than goats, for example. If you plan on getting a horse, cut your fly population first. Horses *really* suffer from flies and whatever you can do in advance of buying a horse will make a big difference.

Fly traps work well also. Either they can be made at home or you can purchase your first one and build more from its design. They are listed in the back pages of most alternative magazines (under "Biological Pest Control"). Make one more than you need to use when changing the others. If you have a trap full of live flies that must be emptied, put your extra trap there until the full trap's flies die (by drowning or starvation), then replace the original trap.

Whitewashing the inside of barns is supposed to reduce the fly population by its brightness. But I found when I had a darker barn the animals often spent the afternoons indoors because the flies wouldn't follow them in. (Whitewashing, however, does offer considerable protection to wood surfaces and is non-toxic; it makes the walls easier to clean and brightens the inside of the barn enough to make a difference to both animal and human eyes).

I suggest having someone else do the job; I have yet to locate a whitewash recipe that isn't a major job to make at home or that contains readily available ingredients. And in every rural community, there is a small company that can do the job in an astonishingly brief time for a low, low price.

Once or twice a year scrub down partitions, mangers, and stall doors with a stiff brush. Scrub feed buckets, waterers, and such once a week.

Use a disinfectant designed for use around dairy animals. Lactic acid in milk has bacteria that are killed by such disinfectants. And since milk picks up everything, these disinfectants are generally safer than other kinds. Most are designed to be used in a heavy water-dilution and more is *not* better. Depending on the brand, it may even be harmful to add more.

Avoid cleaners with phenol compounds. These are toxic to many animals. Use common household bleach in a pinch but it may cause slight deterioration to wood surfaces.

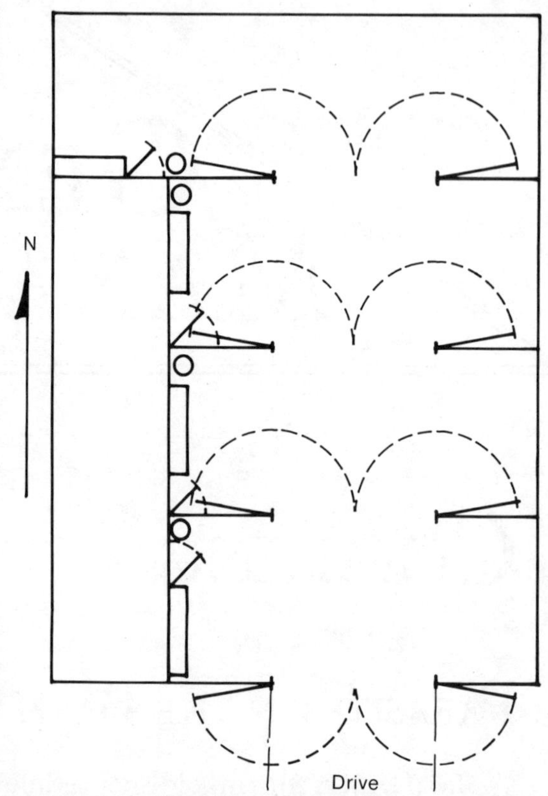

FIGURE 36. FOUR-SPECIES BARN

More on Cleaning

Set up a cleaning schedule that you can live with, and then follow it. The single species manager will be inclined to have *more* disease problems, but your problems will be more complex if allowed to get a good hold.

To make major cleanings easier, large-scale, single-species barns have a central alley between rows of animals. This layout is often copied on a small-scale basis when another system would be much more efficient. Floor and roof space are wasted, not to mention the increase in barn size without any increase in efficiency inside the barn.

One idea would be to arrange all but the last stall at the alley end to open in the center, and put the stall's inhabitants outside while cleaning takes place (see Figure 36). Work backwards from the last stall.

Another way is to have all stalls open to a central location just large enough for a truck or wagon to back into (see Figure 37). This design makes very efficient use of space.

Still another way is to clean from *outside* the barn proper. Close animals into the pen or a portion of the stall. Build gates wide enough to get equipment through into each pen. This saves tedious work with the wheelbarrow in getting wet bedding to the outside and unloaded (see Figure 38).

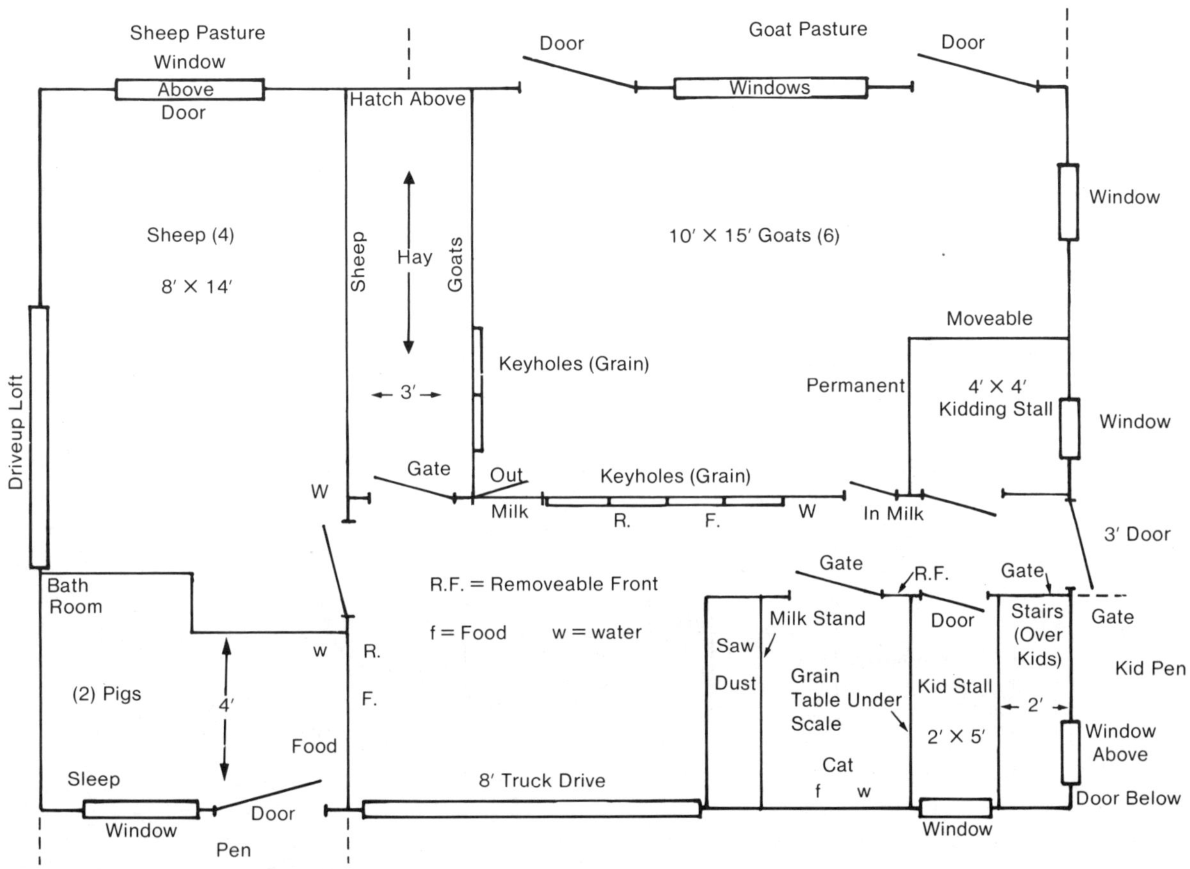

FIGURE 37. 18′ × 26′ BARN

Some sheep barns use a narrow alley for feeding from within the barn; the stalls are arranged where the whole side of the barn opens for cleaning (see Figure 39). This limits the number of species if the straight form is used. Other variations are possible for a wider variety of animals (see Figure 40).

The most important point about cleaning is the potential danger from innocent-looking manure and bedding piles. Parasites grow in them and pass back into your animals unless removed from the area. Whenever you're feeding parasites along with your animals, your feed bills are higher and your production lower. Even a day or two of that manure pile sitting around and there will be a noticeable increase in your parasite problem.

Whenever you completely clean an animal's stall, lime the floor before re-bedding. Use dolomitic or pulverized limestone, *not* agricultural lime, which can be harmful to the animals. Spread immediately after cleaning; it knocks down odor substantially and changes the pH of the floor so pests won't find it comfortable. It also helps kill acid-loving bacteria. I keep a five-gallon pail of it in

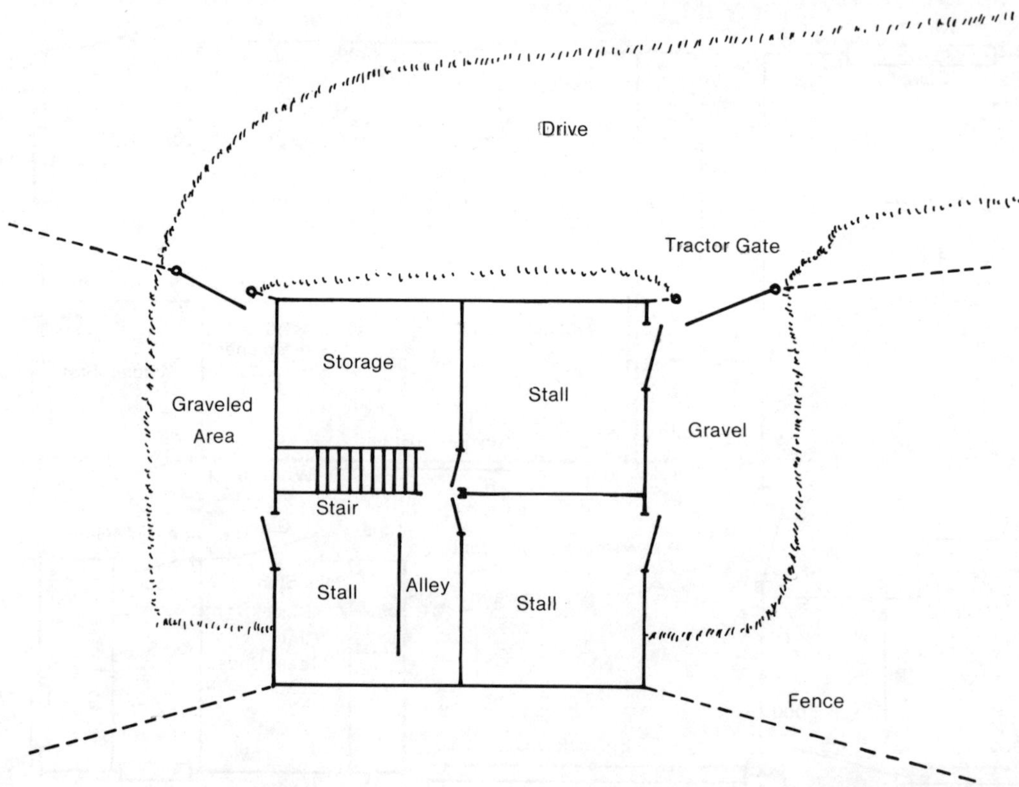

FIGURE 38. CLEANING STALLS FROM OUTSIDE BARN

the barn with a coffee can in it for easy spreading anytime.

Parasites are a more serious problem not only for the organicist, but especially for the mixed-species manager, since manure is the main source of reinfestation. But we have several advantages over the spray, spray, spray people. Natural controls, which sprays kill equally as well as the pests, thrive in the organic barn. Flies can be controlled with pteromalids and birds. Lice can be defeated with organically based rotenone powders; although organic, it is still a poison and should be treated as such.

Pasture and Pen Larvae Control

The best larvae control is pasture rotation. The specific rotation plan is based on the cycle of the particular parasite you're having a problem with; your veterinarian will know the life-cycle of the particular parasite if the problem is getting out of hand. To identify the culprit, have a fecal exam performed on the affected animals. For normal parasite

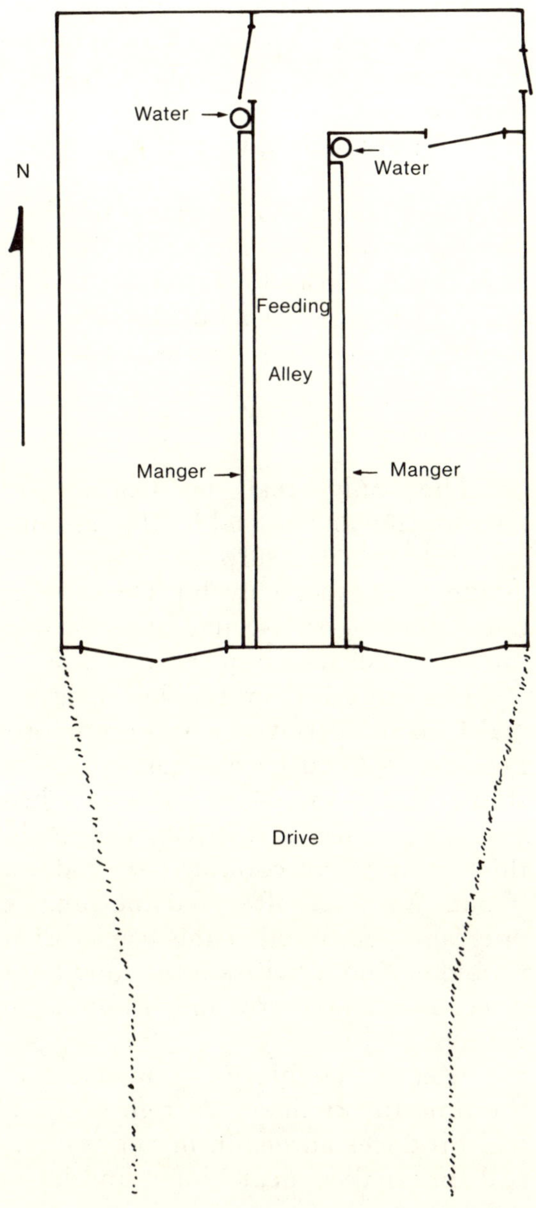

FIGURE 39. SINGLE-SPECIES BARN

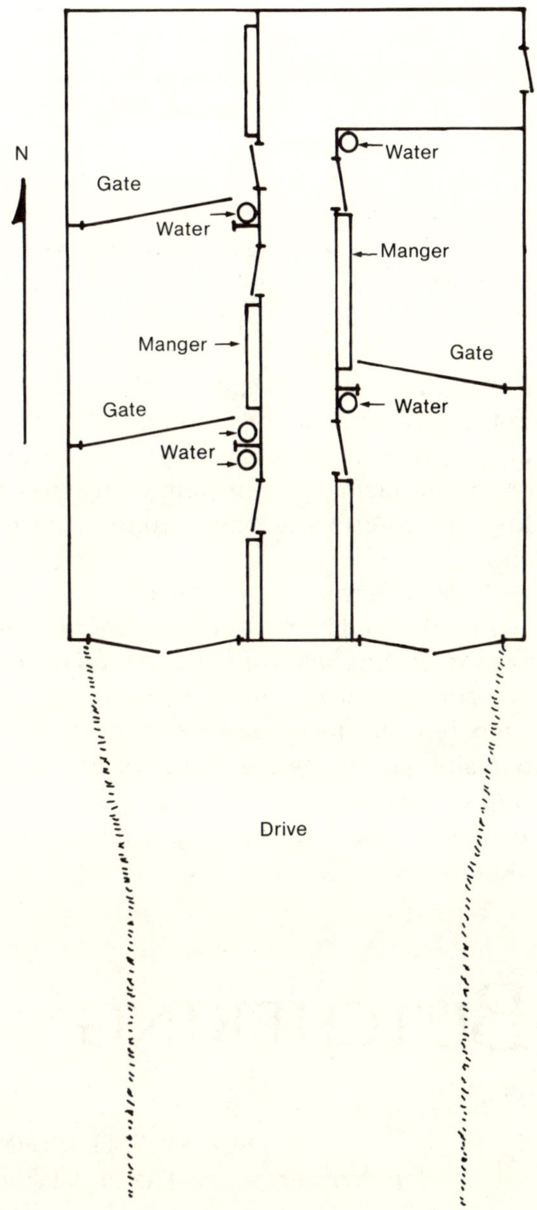

FIGURE 40. FIVE-SPECIES BARN

control, a six-week rotation is sufficient. And it has the added benefit of resting the ground from continuous grazing or browsing.

Periodically sprinkle pens and the barnyard with Borax to cut parasites; this may kill the parasite larvae before they can hatch and reinfest your stock. Borax is the common laundry additive sold in grocery stores everywhere.

As an organic gardener, however, I know trace minerals, such as boron through Borax, are applied to a deficient soil in minute amounts. Any increase in those small amounts can create a toxicity in the soil that may be assimilated by the plants. Borax, sprinkled regularly in an animal's grazing area or pen, could make the plants growing there toxic to the animals.

My own vet (a big fan of chemical farming) thought it would *probably* leach away. (My emphasis.) I tried several other veterinarians with similar results; several even felt the test results of Borax as a parasite control were dubious anyway. Add to this the fact that because Borax is available only in small packages, it's expensive.

BUTCHERING

Natural law has no pity.

ROBERT A. HEINLEIN,
THE NOTEBOOKS OF LAZARUS LONG

For those of us not raised in households with the Sunday ritual of killing and cleaning a chicken for dinner, this butchering business can be pretty tough to take.

If you decide to try it, by all means do! But if you can't bear the thought or if you have wound up flailing an animal to death, give it up for awhile. Sacrifice some of your profits to the butcher. And don't let anybody give you a hard time about it.

If you can watch the local butcher's technique, do. If you can help out after awhile, do. Before long, you'll probably wonder why you pay a butcher when you could so easily do the job youself.

•

The only real justification for slaughtering any animal is, "I am hungry and this animal is surplus." Every other argument is usually invalid. The only real crime is waste, so use all that you can for food or other useful purposes.

Use entrails from butchering as animal feed for carnivores and omnivores. Pest cycles need close attention. Wash stomachs and intestines out with a hose *thoroughly*. Then give everything edible thirty minutes of cooking before doling it out. We have often fed the lungs of one pig to another in this way with no problem. And it helps on the food bill in direct relation to how much butchering you do.

When slaughtering animals, take the time for kindness. A frightened animal produces adrenalin in her system—and that makes meat tough, and lowers quality.

I've stopped withholding food prior to butchering. It makes the animal fret and frighten more easily. It doesn't make an appreciable difference in doing the job—at least not enough to justify the upset to the animal.

There are some simple methods of killing animals that are less traumatic on everyone, including you. Do your killing outside the barn; it unnerves some animals to see (or hear) others killed, and the outdoors is easier to hose down. If flapping carcasses bother you, drop a post maul on the fowl's head while the head is resting on the chopping block. Flapping does help pump blood out of the carcass, but if the job gets postponed because it upsets you, try the maul. There is no flapping at all—nothing.

We used to shoot larger animals, such as hogs, behind the ear. They drop instantly and little blood flows freely when the ventricles are cut. Now we give them a pail of feed directly under where we're going to string them up. When the pig sticks her head in, we shoot her in the forehead. Consciousness never returns but the blood flows more freely when the ventricles are cut. String her up immediately and put a wheelbarrow or bucket under the chest cuts to catch the blood.

Many people kill rabbits by dislocating the neck. The first time we tried it, the rabbit regained consciousness just as I went to cut the head off. Perhaps practice makes perfect, but I'd just as soon skip it anyway.

Now we kill anything we can pick up (except fowl) the same way: hold the animal up by the hind legs with one hand and wham them in the back of the head just once with a piece of firewood or pipe. Make it a home run and hit like you mean it.

One last thing. If you can afford a hand gun, get one. The problems involved in aiming a rifle two feet long accurately can make the difference between a successful butchering and a very messy situation.

Tying and Tethering

The cruelest of cruelties is neglect.

CONFUCIUS

Avoid tying an animal because there are many dangers which are difficult to avoid. The animal that is tethered is subject to hazards she cannot cope with in this relatively helpless state, and she is dependent on you to be aware.

Stray dogs are the number one problem in raising small stock these days. While some species can partially defend themselves, a tether limits defense drastically. And the majority of animals are helpless against dog attack, tied or not.

Water is another problem for the tied animal. When you first tie an animal, they usually aren't thirsty and may overturn their water supply. Or the line may have fouled the water or tipped over the bucket by the time the animal *is* thirsty.

The tethered animal is exposed to anything that comes along, such as the sun. Sunstroke is a serious problem. Put simply, the animal's "inner thermostat" goes wild and the body heat continues to rise until the animal either falls unconscious or dies.

Tangling is also dangerous to your animal. Twisting and binding dangerously shorten the tether in no time. An animal tethered by a collar can strangle very quickly once the tie binds. A halter can force her into standing with her head down which severely cramps neck muscles and blood flow to the head.

A long tether will catch on or get tangled in other objects, holding the animal fast in a totally helpless position. The object doesn't have to be solid either; the tether can catch on a bunch of weeds whose collective root structure holds a very large animal firmly, because her ability to resist is also limited by the tether itself.

Above all, tethers endanger an animal's legs. Legs bound accidentally can be quickly and easily broken or torn, especially if the animal panics when she realizes she's caught. The larger animals have an especially difficult time: if they fall, they may not be able to regain their footing because they cannot shift their weight as they normally do. Horses have the added problem of a solid hoof with small, easily bound bones above, and the tether will not readily expand over the large hoof from the small circle around the pastern.

A broken leg is infinitely harder on a four-legged animal than on us humans. We take it philosophically and know the inconvenience will end in a given length of time. To the animal, proper digestion of food depends on adequate exercise. And you cannot leave her with the herd for fear she cannot protect herself in the pecking order. If she falls, she is helpless to regain her feet without assistance—if she's a big one, *you* too are helpless.

Besides, what can you do with a cow who, because of a broken leg, cannot stand easily to be milked? A herd dog that must hobble after stock? And what about the horse? Modern methods have made it possible to save more horses than ever before—but the ratio is pitifully low. Prevention is still the best medicine!

When Tying Is a Must

In spite of the dangers, there are still times when tying is the only answer. Reduce as many risks to the animal as possible.

1. Put the tied animal where you can check on her often—*then do!* Never leave an animal tied when you must go away, even for a short while.

2. Never use rope. Because it will bind on itself, catch and hold on its own surface, nothing has to get caught in it to put your animal in danger of strangling. Furthermore, rope burns when it catches on anything, especially on an animal's leg. The most common catch point is below the dew claw, the horny appendage(s) just above the foot. Rope burns in this area are a nightmare to heal since the animal must flex the area to walk. Medications wipe off in grass or bedding, especially if it's wet. Then too, the animal steps in manure which makes the area extremely difficult to keep clean. And bandaging restricts healing. Therefore rope is out.

There is one exception. If an animal is tied very short and can only move freely in an arc, as up against a solid wall,

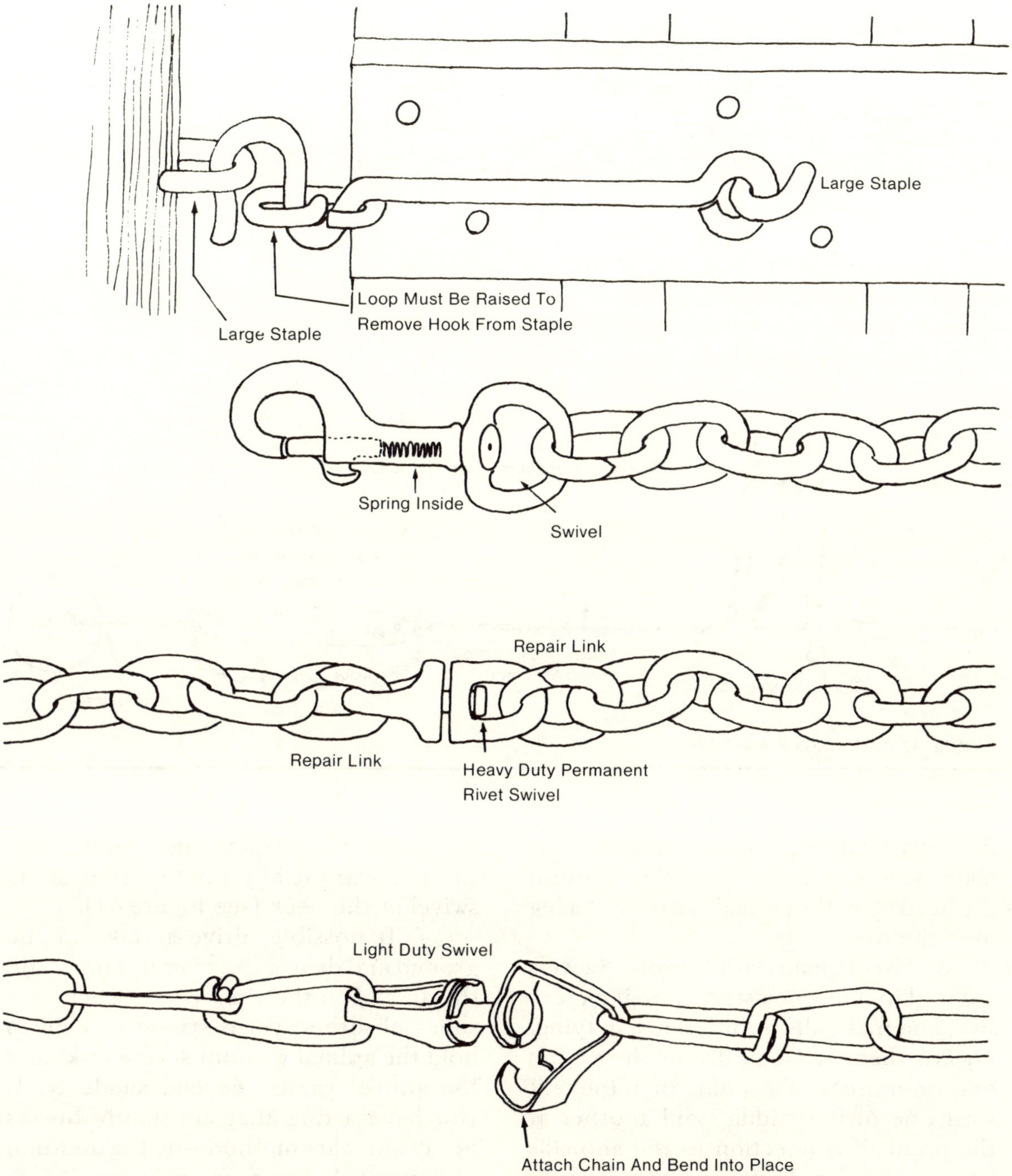

FIGURE 41. DOOR LATCHES

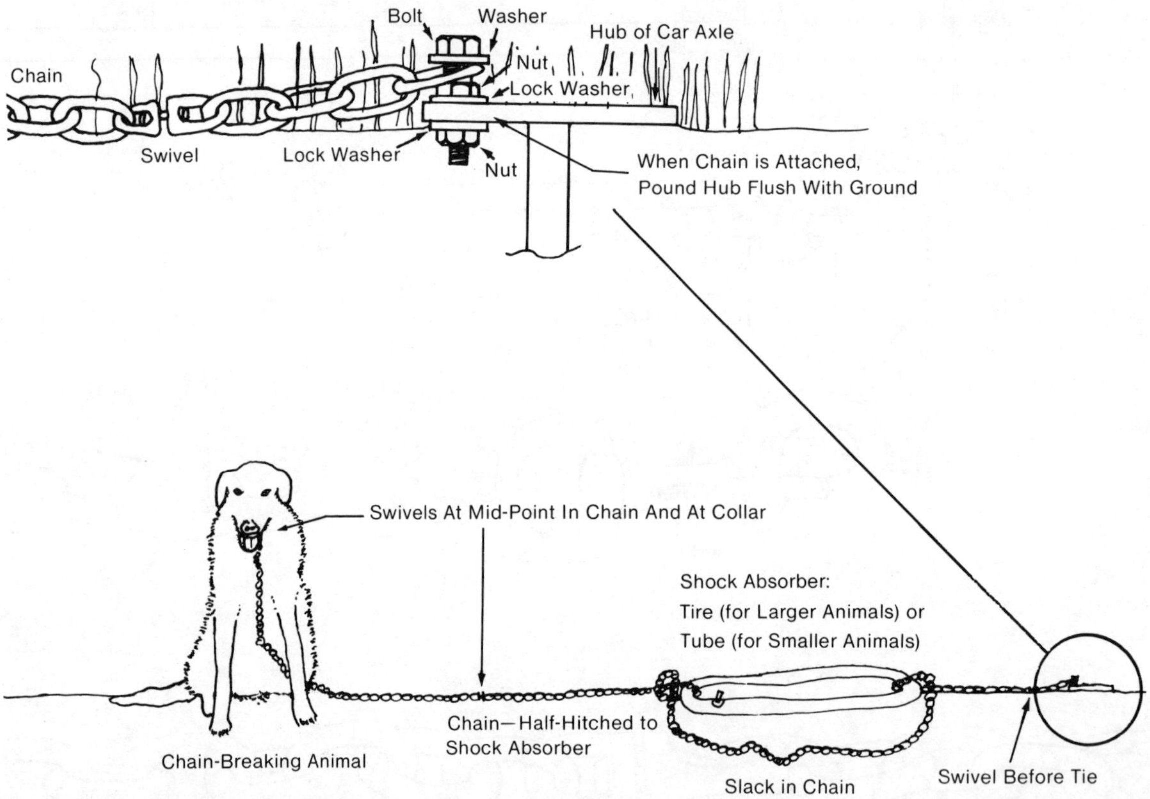

FIGURE 42. CHAINING OF ANIMALS

then it's relatively safe to use rope. But make sure no matter what the position the head is in, the animal cannot get a leg over the rope.

3. Use a chain with swivels. Swivels are sold in most feed stores, are inexpensive, and make all the difference in tying. Locate them strategically in the chain; one or more at the point of tether, at least one in the middle, and another at the point of connection to the animal's halter or collar. In this way, you minimize any possibility of the tether binding. The animal can reverse her position at the end of the tie and the chain will turn over at the ends and middle. Any turning near the point of tie will turn the swivel at the neck (see Figure 41).

4. If possible, drive a stake in the ground and drop a tire over it, then chain the animal to the tire. The tire acts as a shock absorber, yet is strong enough to hold the animal without severe jerking if the animal yanks the end suddenly. If you have a dog that constantly breaks her chain, this method—or the addition of any shock absorber—will stop it (see Figure 42).

If the animal becomes wise to slipping the tire off the stake, use a stake with a wider top and tie or chain the tire to it.

Many people have simply chained an animal to a heavy weight with success. But a neighbor once tied a horse to a brand-new, expensive tractor tire with dire results. Feeling confident that the horse was secure, he left home. When he returned, horse and tire were gone. The horse turned up at a neighbor's, but the tire never did.

5. Tie the animal in a clearing within reach of a shade tree; the chain should fall short of the trunk. The animal then has a choice of sun or shade.

6. At the chain's greatest distance from the stake, set a water bucket in an old tire. It's not easy to tip a bucket inside a tire.

7. With heavier animals, tie for a quick release. If an animal does fall and she's too big for you to move alone, a tie that can be undone under tension will make the difference. She can then be freed to regain her feet under her own power.

8. A general rule of thumb: "tie low and long, tie short and high." A long chain up high will hang like a jump rope. It may catch the animal's legs or her neck as she lifts her head up.

9. Never tie an animal near an object (such as a fence) she can jump. In panic, she may try to jump.

10. Don't forget to check her often!

A few other items should be noted here. Rope and nylon halters or collars cause special problems. Some brands of rope halters shrink in the rain. Nylon, under stress, shears off hair and, if the stress is increased may shear off skin. Check ALL halters and collars regularly. On at least one occasion, in the case of a young colt I purchased, I had to surgically remove the halter from under the skin that had grown over it.

If an accident occurs and you find an animal suffering from sunstroke, lower her body temperature as quickly as possible. Put a small animal in a cool, dark place and watch over her. Spray a large animal with a garden hose to dissipate heat faster. Any animal suffering from sunstroke is under great stress, so take care, and avoid handling as much as possible. Keep them cool, quiet and, if possible, in the dark.

Dog collars are a special problem. If your dog has a choke collar and license, be sure the license is on the *pull* ring and *not* the stationary ring. As the chain slides through the stationary ring, it can catch on the license and wedge it across the ring, leaving the dog with the chain jammed at maximum tightness. This could cause slow strangulation.

Pasture Management

The best way to assure a wide range of vitamins and minerals is to feed a wide range of plants and grains. The most practical way to do this is by allowing the animals to forage for food on pastures.

Gene Logsdon

Differences in location of land, soil composition, soil acidity, climate, and past and present management make generalities on pasture management difficult. But a few facts will help you begin to manage and improve what you have.

As a general rule, unimproved pasture will support one head of livestock (cow or horse) per *five acres*. Smaller species (such as sheep or goats) equal that cow or horse in groups of three to seven animals. To use a median figure, call it five sheep to one head of cow. The variation depends on what volunteer crops (natural plants and grasses growing without specific planting) are growing on your land. On a theoretical ten-acre field of volunteer crops, expect (not bet) to support two cows or ten head of sheep.

Of course, you'll know the field isn't providing enough food if your stock begin losing weight. The early signs of insufficient pasturage are important to watch for, because you never want the situation to get as advanced as weight loss. Watch for stock walking back and forth over the same strip of ground examining the grass often. Other signs include: grass that is eaten below six inches of growth, a "clumpish" look to the grass, or any visible plant roots. You may be able to make do by supplementing their diet with hay.

There are a few inexpensive things that can be done to better manage your fields and pastures.

Let's continue with that ten-acre field with the two cows. You've already improved the ground somewhat just by letting the stock out on it. Grazing improves the soil because the cows will fertilize it as they graze. To get the most use from that ten acres, fence it into three strips or sections, then rotate the cows on the strips. The ungrazed portions get a rest and this cuts down on the parasites and eliminates the erosion caused by paths.

Mowing the pasture from time to time reduces the number of weeds reaching the seeding state and prevents plants from becoming woody and tough. It improves the root structure of the better plants, and the new growth will be more palatable and nutritious for your stock. Mowing also invites new plants to seed since they are no longer competing with older, more established weeds.

Liming a field can be done on a scale as small as a pick-up load (or a few sacks of lime) at a time. For $25, a pick-up load of pulverized limestone will vastly improve your pasture. (You'll get agricultural lime if you don't request pulverized limestone. Pulverized limestone is cheaper and won't "burn" plants and worms the way agricultural lime will.) Changing the soil's pH will noticeably change the composite of volunteer growths. Plants like clover will begin to spring up, even without planting it. Do a small portion well rather than a large portion of the field poorly.

Improve alkaline land by adding organic matter, especially manure. It too will change the types of growing plants. Again, it's better to improve a small portion a lot than a large portion a little.

Many types of pasture seed can be purchased at minimal expense. Do as large a portion as your finances allow. Make up in quality of seed-bed preparation what you can't do in finance or size.

Mow your newly planted area as the need arises to help thicken the roots and to keep the growth from becoming weedy.

For further and more specific information, remember your Soil Conservation Service and your Cooperative Extension Service. One or both are located in just about every county seat in the nation. Their services are paid for by your tax dollars. They have the most specific information available on a county basis. They will provide you with free advice for your specific property. The Soil Conservation Service even supplies soil survey maps and photos at a relatively low cost; use these to plan your land-use more accurately.

Pasture improvement can do truly remarkable things. Given sufficient management, you can make that theoretical ten-acre field actually reverse the earlier figures to *five* cows per *one* acre. And that can change your whole livestock program. If you've felt you could use more land than you have, pasture improvement might be the way to get it—without ever leaving the farm.

Electric Fencing

Fencing is important no matter how small your livestock holdings, and it's darned expensive stuff.

There are two new types of electric fencing: the Gallagher and New Zealand style Spring-Tight Fencers. They are both energy-efficient and keep stock in. They run the greatest distance with the maximum charge and are effective even when run through wet weeds. You can obtain further information about either or both types by writing to:

Henry and Cornelia Swayze
Brookside Farm
Tunbridge, Vermont 05077

They will send you information on request and include a list of dealers in your area.

Record Keeping

When there are only a few animals of a few species, I keep records on a calendar hung in the barn on a nail, along with a pen. When the population is greater, I carry a notebook, and milk charts, in which I note anything unusual. I also date each entry. Either way, records are extremely important in management. If an animal escapes on Tuesday, you might have forgotten the incident on Friday when you notice she's constipated or scouring. Records also provide you with answers when the vet asks, "Did you ever notice this before? How long ago? What was the behavior previous to the incident?"

Bigger patterns both in your management and the animals are discovered easily from records. They also are very

SEPTEMBER	3RD	4TH	5TH	6TH	7TH	8TH	9TH	WEEKLY TOTALS
BELLE	A.M. / DAY'S TOTAL IN POUNDS / P.M.	A.M. / DAY'S TOTAL / P.M.	A.M. / DAY'S TOTAL / P.M.	A.M. / DAY'S TOTAL / P.M.	A.M. / DAY'S TOTAL / P.M.	A.M. / DAY'S TOTAL / P.M.	A.M. / DAY'S TOTAL IN POUNDS / P.M	WEEK'S TOTAL / AVG./WK / TOTAL LACTATION TO DATE
PIPPIN	3¼ / 6¼ / 3	3¼ / 6¾ / 3½	3 / 6 / 3	3 / 6 / 3	3 / 6 / 3	3¼ / 6½ / 3¼	3¼ / 6¼ / 3	43.75 / 6.25 / 1,635.25
RUBY	4 / 7½ / 3½	4 / 8 / 4	4 / 7½ / 3½	4 / 7¾ / 3¾	4 / 8 / 4	3½ / 7 / 3½	3¼ / 6½ / 3¼	52.25 / 7.5 / 2,014.75
TINA (FIRST LACTATION)	2¾ / 5¾ / 3	3 / 6 / 3	3¼ / 6¼ / 3	3 / 6 / 3	2¾ / 5¾ / 3	2½ / 5 / 2½	2¾ / 5¼ / 2½	40.0 / 5.5+ / 1,246.25
SETH-ACTIVE	CLOUDY	SUNNY	SUNNY	RAIN	CLOUDY	SUNNY	SUNNY	
	TRIMMED	A.M. FED	PIPPINS	PIPPIN/	FED CORN	FED CORN	MAINTENANCE	
	HOOFS/ALL	COMFREY	APPETITE	SEASON	STALKS/ALL	STALKS/ALL	CLIP/ALL	
			DOWN	BRED 2X	TINA APPETITE	TINA APPETITE	TINA SEASON	
					DOWN	DOWN	BRED 2X	

FIGURE 43.

FIGURE 43A.

helpful in culling, feeding-change experiments, and patterns in breeding and birthing. And they give you a record of exactly what day you trimmed hoofs, clipped goats, sheared sheep, or bred your animals. These are very helpful to your future management; they also let you check back on yourself to see how far you've come in your abilities. You can improve without records, but you can't know by *how much* without them.

Using your own daily information chart, you will note what additional food supplements (recorded at the bottom of the chart) effect milk production fluctuation, how the weather affects various milkers, and the possible causes of scours. A decrease in milk production, as well as a decrease in appetite in an individual frequently indicates a doe's oncoming season.

STRUCTURE

It was a saying of Bion, that though the boys throw stones at frogs in sport, yet the frogs do not die in sport but in earnest.

PLUTARCH

BARN LOCATION

The ideal location of your barn is one point of a triangle with the house and garden being the other two points. This layout makes it easy to go from the barn to the house for chores, bring goodies from the garden to the animals, and haul manure from the barn to the garden. Given a choice, I'd locate a barn uphill from the garden; it's easier to haul manure downhill than up. Somewhere in the middle, locate the compost heap.

Of course this is an ideal. My own barn is downhill from my garden, but I was able to retain a triangular pattern. If I go to the garden in the early spring to check on the strawberry plants, and find the comfrey up, I'll cut a little and take it to the goats on my way to the house. I've located my compost more or less behind the house and uphill; it means household contributions must be carried uphill. But big loads of compost for spreading are on a level with the garden. Garden produce comes downhill to the house. One adjusts as necessary to site, but ideals are nice when they can be had.

Wherever possible, have a driveway that serves both house and barn. Driving over the lawn with a pickup full of hay or grain supplies makes for messy lawns.

Most folks build rectangular barns. It's a convenient shape for housing animals, and the most common. But give some thought to which way to build that rectangle in advance.

The windbreak effect on the housing can be substantial just by the way you plant that rectangular frame. On the average site (if there is such a thing) the major source of weather is from the north, seconded by the northwest. Putting the small end of that rectangle to the north will keep your barn from being broadside to those prevailing winds. It also puts your opposing small end to the south, which some people object to.

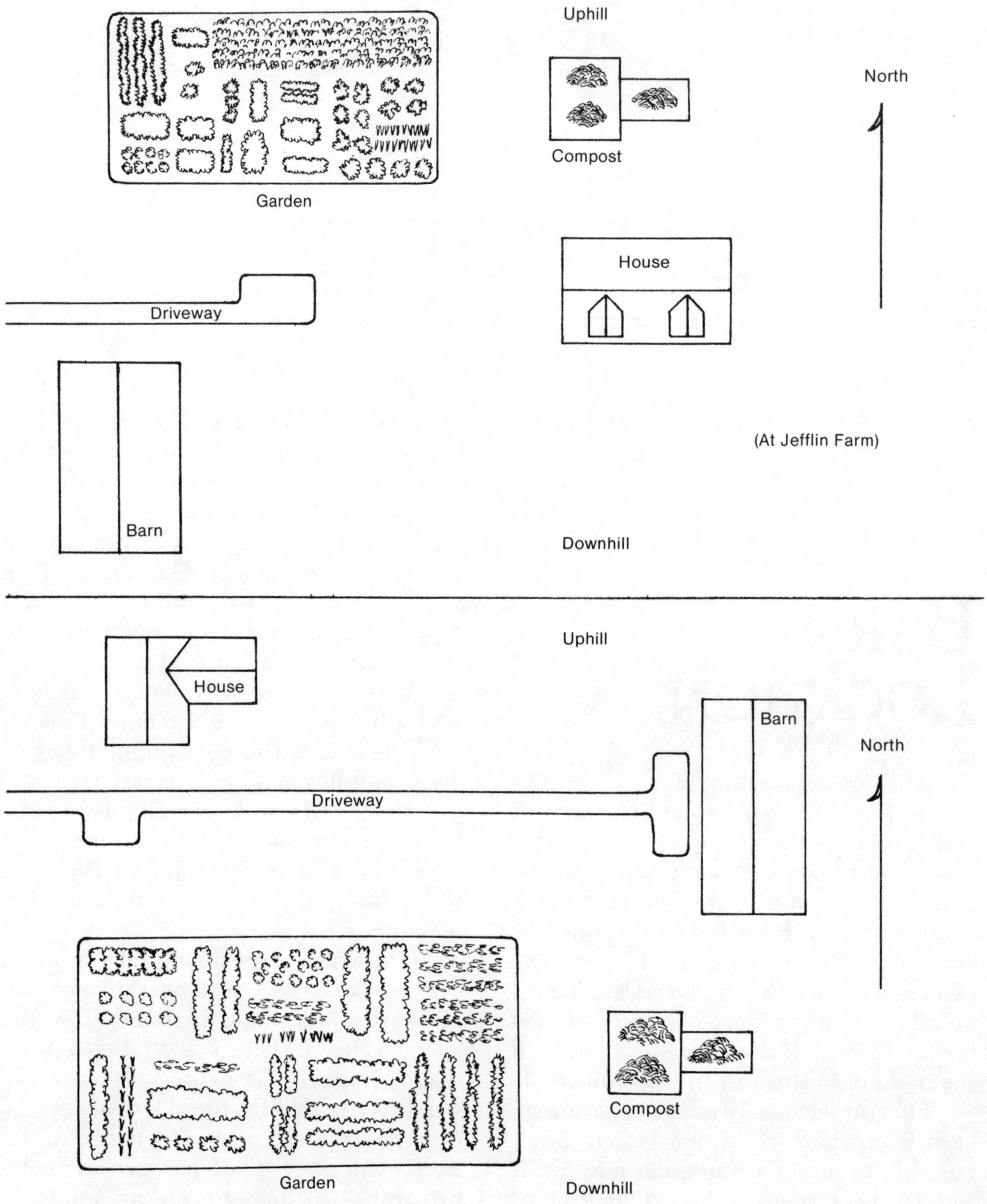

FIGURE 44. TWO POSSIBLE FUNCTIONAL YARD PLANS

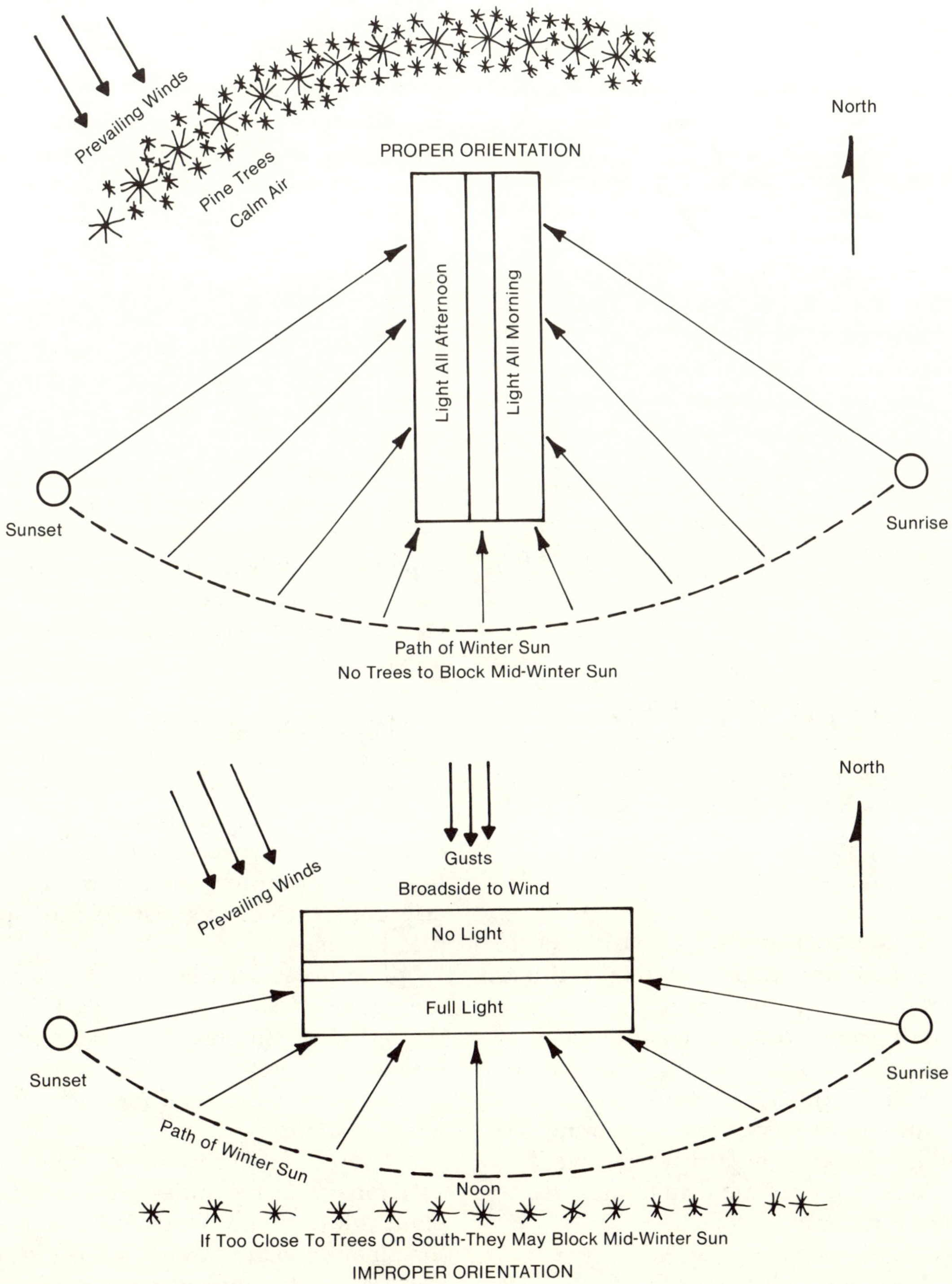
Prevailing Winds
Pine Trees
Calm Air
North
PROPER ORIENTATION
Light All Afternoon
Light All Morning
Sunset
Sunrise
Path of Winter Sun
No Trees to Block Mid-Winter Sun
North
Gusts
Broadside to Wind
Prevailing Winds
No Light
Full Light
Sunset
Sunrise
Path of Winter Sun
Noon
If Too Close To Trees On South-They May Block Mid-Winter Sun
IMPROPER ORIENTATION

FIGURE 45.

Although the south side is your best source of sunlight year-round, it only benefits those animals located directly against the south wall. Those on the north side would have the winds stealing their body heat and no sunlight for comfort. And it would be too hot in the warmer months for those on the south.

If your housing runs with the long sides facing the east and the west, stalls inside running along those walls are guaranteed some sunlight every day it's available. And those needing the most sunlight may be located farther towards the south. Although there is little difference in the sides of the barn in such an arrangement, locate those animals that like it a little warmer on the east side of the barn where there is the least possible weather influence.

VARIETY OF CONSTRUCTION TYPES

CONCRETE BLOCK OR STONE BARNS are more common than you might think. They're quite permanent, which makes it hard to change your mind about location of doorways. They can be damp if not perfectly ventilated. They are time-consuming and supply-consuming to build, making them fairly expensive; but if you have more time than money, stone may be the way to go.

LOG CONSTRUCTION is ideal for those who have plenty of time and nearby trees. To build a barn out of logs demands a lot of wood, time, and skill, and they are hard to make draft-free unless you're using prefabricated logs. It's rare to find an old log house or barn in good condition, since logs begin deteriorating the moment you finish the structure. And they're ecologically wasteful because they take lots more wood than is really necessary.

PLANK AND BEAM is an old type of structure which is growing in popularity. It too takes a great deal of wood.

WOOD FRAME is also a waste of wood in a barn, generally speaking. If you leave the frame open to the animals on the inside, there'll be more wear and tear on it; if you cover the inside walls, you create rodent alleyways and use too much wood.

POLE CONSTRUCTION was originally designed for farm use because it's the most economical and simplest type of construction. It has the highest wind resistance of any wood structure and will hold larger amounts of hay when properly constructed. It's fast to put up—a nice bonus.

The poles are the major structural members. Most poles will last longer than you will if installed properly. Although changing a pole sounds truly horrendous, it can be done without too much trouble.

There are three basic substances used in treating poles commercially: pentachlorophenol, chromated copper (or zinc) arsenate, and creosote. If you want a specific kind of treated pole, *you must ask for it*. Pentachlorophenol has been known to make animals sick. I've never heard anything about arsenate treatments, I admit, but creosote is still my choice.

Creosote-treated poles have been used around livestock and people for a long time without difficulty. Creosote is a natural substance derived from wood. You cannot paint a creosoted pole, which is the major objection for some people.

A word of caution: if you chip, drill or saw open a treated pole, treat the opening lavishly with wood preservative to prevent rot from entering at that point.

Installation of the poles is the most important part of your structure. If you combine a concrete floor with a pole structure, pre-treat the poles from the ground level to twelve inches above with tar. Further, cover *loosely* with burlap and coat heavily with another application of tar. In theory, concrete will not crack from expansion/contraction pressures around a circle. However, dried tar on burlap will provide flexibility. Slightly bevel the concrete away from the pole so water won't collect on the pole.

Foundations properly built and drained make more difference in the life of a building than any other single factor.

Lumber Considerations

Generally speaking, the best type of lumber to use for your barn is whatever you can most easily obtain. Local lumber made from the more common type of tree in your area is usually cheapest because it's readily available; and if you order in quantity, your local supplier will probably deliver it free of charge.

Beyond these considerations, there are other factors to assess.

Hardwoods

Hardwoods like maple, oak, and ash are rarely used in construction anymore because they are difficult to work with. The density of these woods makes nail driving a serious chore, saws dull faster and, once seasoned out, driving a nail back into a loose board may mean drilling a hole, as the density increases with age. Hardwoods also tend to dry unevenly, and may warp because of it.

Hardwoods have benefits, too. They can stand the abuse of dampness, animals and weather longer than other woods. And they have a greater compression strength, but are less flexible than softer woods.

Softwoods

Softwoods are more common in construction. They are easy to work with, and even drying and seasoning make them extremely popular. Of the softwoods pines and soft maples are the most common. Three years after the barn is constructed, you can replace a loose board easily as the wood remains somewhat porous. Its porosity can be a benefit: it will take paint, whitewash or preservative much better. An untreated surface, however, will absorb and lose moisture from within or without the barn more frequently than hardwoods—which means a somewhat shorter life-span if the surfaces are left bare.

In our area, hemlock is the most commonly available wood. It's like a semi-soft wood. It's easy to work with but dries very hard and evenly. Although it's difficult, you can drive a nail long after the wood has seasoned out. It's shown considerable strength in every use we've put it to, including our own house construction. It comes down once again to what's available.

Kiln-Dried or Green Lumber?

Kiln-dried lumber is consistent, warp-free (if the drying has been done properly) and will not shrink after the structure has been standing a while. It will accept and hold nails with equal strength everywhere.

Architects have a lot to say about using green wood in construction. They say it warps and that shrinkage after construction causes basic structural weaknesses. But architects build on other people's money. They also contract out for the actual labor and need to know quality and consistency will result from the average carpenter's work.

People constructing their own barns or homes, however, are usually doing it on out-of-pocket expense accounts. And that means another factor must be considered—*time*.

Construction under the latter conditions is a much slower operation; the wood will either be seasoned in the pile awaiting use or be nailed straight and true in place, seasoning in the open air long before the structure is put to use. (Remember the hardwoods dry less evenly.) When building is done on weekends and holidays or evenings after work, the entire structure is seasoning and settling as it's put up.

Then there are problems with kiln-dried lumber. It's expensive. And the quality of today's kiln-dried lumber isn't what it ought to be; because the lumber is dried to below normal moisture, the boards take up moisture in storage. Consequently, you often find yourself with shrinking kiln-dried lumber once it's in place.

Green lumber does shrink. Cut squarely and nailed properly in place—always nail an inch in from *all* edges—green lumber will shrink in place as construction continues. This prevents warping, and shrinkage "problems" won't occur because the shrinking will be uniform.

If the issue concerns you, dry your own rough-cut lumber in advance and have the best of both worlds. In drying your own (see Figure 74) it is important to turn the top board at least once a day. Put in sufficient supports to keep the pile from sagging in the middle and turn the whole pile, one board at a time, once a week. The lumber will dry evenly and straight. Dry it under a roof if possible.

Planed or Rough-Cut

Planed lumber is more expensive. It's somewhat easier to work with and has fewer small splinters on the surface. Its dimensions are most often consistent and lengths are standard and precise. But the dimensions are smaller than those of rough-cut lumber: that is, a 2 × 4 planed is really 1½ × 3½.

Rough-cut lumber is full-sized—a 2 × 4 is really 2 × 4. The surface is splintery, so you must be careful.

It pays to be selective if you have a choice of mills from which to buy rough-cut. Take your tape measure to the mills and check for *consistency* in width over the length of a given piece of lumber. The more accurate the dimension, the easier the building. This does not apply to length, as longer-than-stated length may be beneficial if you're building where the length is longer than standard (such as spanning a length of eight feet, two inches). In such a situation with planed, sized-lumber, you would have to buy a ten-foot piece and cut it down to eight feet, two inches. With rough cut, the eight-foot piece will be long enough.

Lumber Lengths

Loggers often take logs from trees in eight-, ten-, and twelve-foot lengths. These lengths are more consistent in circumference, therefore bringing a better price at the mill. They are easier to load and haul to the mill and are easier for the mill to handle.

Because of this, lumber lengths of twelve feet and under are more economical. Conversely, lumber that is more than twelve feet in length not only costs more per running foot, but is in a new pricing category per board foot. (Lumber sizes are always in two-foot increments: 8,10,12,14,16, etc.) While this cost jump is not a great deal higher, it may make a difference on some of your decisions in building. Use of longer lumber will increase the expense in direct relation to how much of it you use.

This is the reason why so many homes built by contractors are sized in one dimension on a multiple of twelve (twenty-four by whatever). That is, a living room may be twelve by eighteen, twelve by twenty-four, etc. The room corresponding to the living room will also be eleven or twelve feet in one dimension. This twelve-foot dimension is determined by the direction of the ceiling rafters above and the floor joists below. The other dimension is free of that limitation because the strength of construction is met by the shorter (twelve-foot) span. In fact, boards of varied lengths make a stronger span between rafters or joists; one of the key effects of their structural strength is keeping joists or rafters from shifting to a new position or spacing.

Twelve feet or less on a given span means better support. A great weight, like a loft full of hay in the center of a fourteen-foot (or more) span may make the ceiling sag over a period of time. Under supports (joists) would have to be increased in width dimension to compensate for the length of the span. An example would be a span of twelve feet being supported by a 2 × 10, or a fourteen-foot span being supported by a 2 × 12. Another way to increase the strength of a span is to keep the joists closer together; instead of being spaced

24 inches on center line, 16 inches on center will make it stronger. Both may have to be done if you increase spans by much or expect to load a lot of hay on top of them.

This is not to say you cannot span great lengths without support posts underneath. It can be done and often is. There are examples in every community, such as indoor riding arenas and some lumber mills. But either your costs will be higher to have someone else build it *or* you will need the necessary skills to build trusses. Trusses support the weight from above—which can get a little tricky for your average backyard builder. And a loft is impractical with the truss supports passing through the middle of the space.

Spans of more than twelve feet (or so) are not usually necessary in small-scale animal housing. Fasten your partitions to support posts; they'll increase the strength of your loft if installed properly.

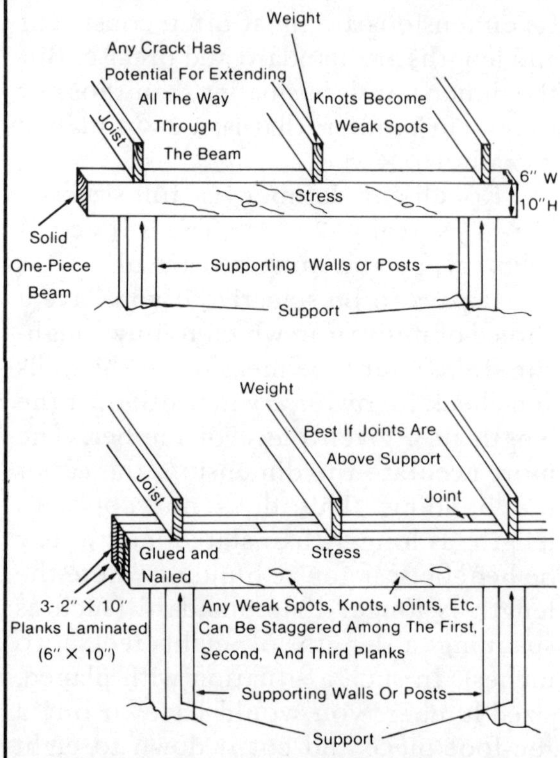

FIGURE 46. LAMINATED VERSUS ONE-PIECE GIRDER or "MAIN BEAM" (6" X 10" Example)

Beams

A beam is any piece of wood that holds one side of a building to the other without a break in the span. Your joists don't fall into this category because there are two (or more) from one side to the other. In fact, they rest on the main beam as their support.

When strolling through old barns, everyone has been impressed by the sheer size of the beams spanning the structures. Some are truly impressive; there is an old barn near here with a main beam twenty-four inches by twenty-four inches. While its size is impressive, so is the crack through its center along the grain of the wood. So just how strong can that solid beam be?

If we've learned anything over the years about construction, it's been the value of the laminated beam. Beams constructed of more than one piece of wood are stronger than those made of a single piece. The variations in grains in the different pieces of wood making up the beam add considerable strength. And the varying lengths stagger any possible stress points (see Figure 46).

For example, a beam spanning thirty feet would be made up of two-inch thick boards, two 15-foot, three 10-foot boards, and three boards of 12-, 12-, and 8-foot lengths. By varying the joint ends so no two correspond, such a beam is stronger than a solid six-inch beam of a single piece of wood, even one with exceptionally straight grain. Nailing them together makes them one beam.

Of course, two or more supports under such a beam would prevent it from sagging for many years to come under stressed loading. The thickness is decided by other factors, such as the type of construction and the weight you plan to hold up with it.

Flooring

Soil

Because it's already there, soil is the cheapest flooring. It has other benefits as well: it's softer to walk on and it's usually not slippery to you or the animals. Soil is a particularly good flooring for horses because the flexible surface keeps their feet and legs healthier. Goats that were improperly bedded have been known to develop keel and knee problems on harder-than-soil floors; in such cases they do better on soil floors.

Unfortunately, in many parts of the country, a soil floor means a mud floor. When it freezes in winter, you have "winter concrete." Of course, it doesn't freeze smoothly, either; instead, it freezes in the shape of hoof and footprints, making walking uncomfortable at best until you once again have the inevitable mud. Stalls must be bedded extra well to keep animals from becoming filthy from the mud. It's impossible to disinfect mud when the need arises and the contamination level is high.

Sand or Gravel

Sand or gravel keeps the animals up off the mud for awhile. In sinking, which it inevitably does, it improves soil drainage somewhat. Sand on the soil surface of a barn floor holds urine instead of draining it. Neither stays where you put it; they are either worked into the soil or formed into holes and ruts on regular animal pathways. They're expensive and, at best, difficult to clean.

Wood

Wood is a pretty expensive way to go, both initially and in maintenance costs. The major drawback is its absorbency; it holds urine and odor like nothing else. Soft wood rots with impressive speed. Hard wood holds up much longer but continues during that time to hold odor and bacteria.

Another serious problem with wood flooring is that it's slippery when wet. Even rough cut flooring is slick when wet to humans, and anathema to hooved animals. Boards take quite a while to dry out once wet, and you have to cope with a small spot of water longer than you care to. A slip could mean a broken leg, injured udder, ruined knee, or abortion. And last but not least, parasites and rodents find the underside of wood flooring the ideal environment in which to raise families or to travel from one place to another.

There are ways around some of these problems. Use mats made of rubber or other good-traction material in the more common pathways for safer walking. Keeping a bucket of sand handy to sprinkle on wet spots helps. And wood floors can be designed for elevated areas, such as for goats in special circumstances.

Elevated Flooring

If your barn has ceilings which are too high for your goats, it's possible to raise the goats instead of lowering the ceiling. Goats love to climb on anything, and seem to enjoy such a location. Install elevated flooring so that the *maximum* space between boards is three-quarters of an inch. This can be attained in two ways: if using dried lumber, use a three-fourth-inch spacer as a gauge to keep a consistent gap when nailing the flooring down. If using green lumber, use a gauge spacer of about three-eights of an inch and keep the goats off the flooring for two weeks. The boards will not thoroughly dry for three months, but in two weeks' time, the surface drying will have taken place sufficiently for you to oil the boards. Be stingy: too much oil will make the surface slick for quite awhile. A small amount will soak into the dried surface quickly. After a day or so, the goats can be put on it (see Figure 47). The lumber

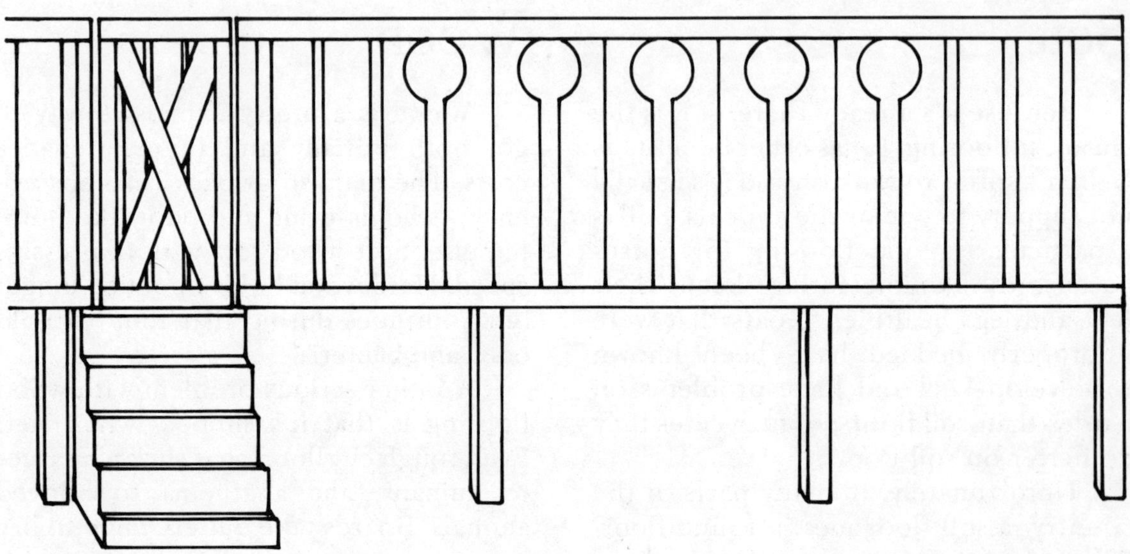

FIGURE 47. ELEVATED FLOORING AREA FOR GOATS

will continue to dry over a period of time, but will do so only on the ends and bottom side.

Raised flooring is for those with special housing problems. It is *not* the ideal; drafts can be a serious problem in elevated housing; some goats will climb halfway out and get hung up. Access in an emergency is essential. Also avoid planed lumber; it *will* cause slipping accidents.

Manure should fall through the cracks, right along with urine, keeping the elevated floor fairly clean and dry. From time to time, you'll have to get up in there with a sidewalk scraper and poke collected manure down through. Major cleanings should be done from underneath with a shovel and rake.

Asphalt

Asphalt isn't slick when wet and provides a resilient but impermanent flooring. The normal period for it to stay in good condition is ten years. The cost of installation is astronomical.

Concrete

Concrete is the most permanent flooring. Its surface is consistent year in and year out. It can be disinfected and is relatively inexpensive to install.

Installation is the key to success with a concrete floor, so put some thought into it in advance. Pour the concrete on a well-drained gravel bed. Wire reinforcing brings up the cost initially, but it's worthwhile. Concrete has strong compression strength but it is weak on flexibility; reinforcing wire keeps it in place if it cracks. For small barns, install the concrete floor with a slope of one inch every four or five feet.* In barns wider than twelve feet, make the center the high point, again with a one-inch fall for every four or five feet. Deliberately roughen the texture of the finished surface, such as with a broom, to increase traction when wet.

Use treated lumber wherever the wood attaches directly to concrete, such as at the sill plate. It's worth the expense.

Concrete, especially when wet, is slippery to hooved animals, although not nearly as bad as wood flooring. Keep some sand handy in a bucket to sprinkle on wet spots. Bedding is also important on concrete. Besides giving the animal a more flexible surface to walk and lie on, the bedding absorbs urine. Urine is higher in nitrogen than manure and is good for your compost or fields.

Concrete is not waterproof. If you don't drain the site well, water can work up through the concrete from below and keep your barn flooded. A well-drained barn site is a must.

Drainage

All barn floors should have drains installed near the water source. These

*One inch fall every four or five feet is a very mild slope.

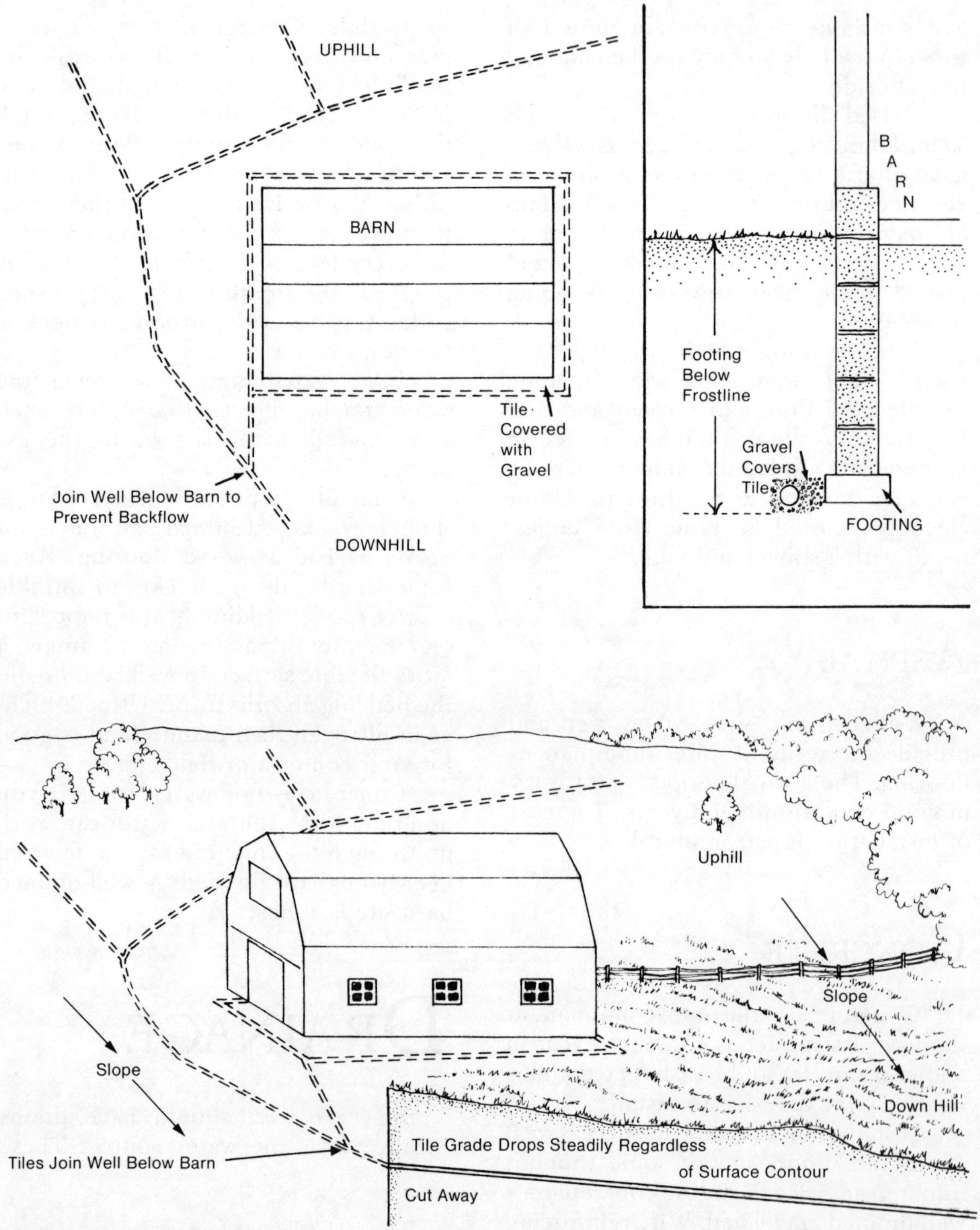

FIGURE 48. A LAND-DRAINING SYSTEM AROUND THE BARN AND YARD

drains should lead to a gravel bed below the floor. General drains in a barn floor usually clog or let water *in* instead of out.

The most important drain system is in the site itself, not only to take water away from the site but to prevent its coming in from outside sources.

Locate your barn in a well-drained area. Barns located on hillsides will flood in the spring thaw unless proper drainage is provided (see Figure 48).

Agricultural drain tile is relatively inexpensive, easy to install, and works exceptionally well. Today it is usually made of plastic and sells for under one hundred dollars for a 250-foot roll. Installation is the major expense. Dig a trench surrounding your barn *perimeter* below the frost line (2½ to 3 feet in our part of the Northeast) on a *level*. Cover the bottom lightly with gravel, set the drain tile in place and cover with more gravel. Dig an *outlet* trench away from the *perimeter* trench, sloping this trench gradually to a lower point (ideally, a creek or watercourse) away from your barn site. Set the drain tile in the same way in this trench. The *average* drop of this outlet is 1 foot in 100 feet of length, and the constant gradual slope is attained by the bottom of the trench, not necessarily the land's slope. All of this is backfilled, where it becomes a silent and highly functional partner.

If your site is very wet, run two tiles side by side, or use a V tile trench (see "optional uphill drain" in Figure 48) above the site as well as the perimeter installation.

Water finds its way into the buried tiles, and any area where they're properly installed will be quite dry. If you need further information on drain tile and its installation, contact your local Soil Conservation Service. They are paid for with your tax dollars, and will give you much needed assistance, usually free of charge.

LOFTS

Years ago, most small-scale projects were handled by sheds and coops to separate different projects—pigs in one shed, wood in another, chickens in a third and so on. These were small buildings, built as the need arose and allowed to deteriorate as the need waned.

Today, we combine housing whenever possible, but an astonishing number of people build barns with no lofts. I began asking them why a few years ago and the general response was to keep initial costs down; they don't grow their own hay anyway, so what's the point? Hay can usually be purchased less expensively in large lots, so for many, it seemed easier to pick it up once a week at the neighbor's and "store" it at home on the barn's first floor (since there is no second floor). Those who didn't add a loft at building time are now realizing that the cost of gas to pick up hay once a week over a six month period is more than enough to pay for the loft.

The potential contamination of hay on the first floor is a problem, too. Children play on it, fouling the hay with manure from their shoes. Dogs and cats may feel a need to "mark their territory" on it (although I've never seen either species do that in the loft) and if an

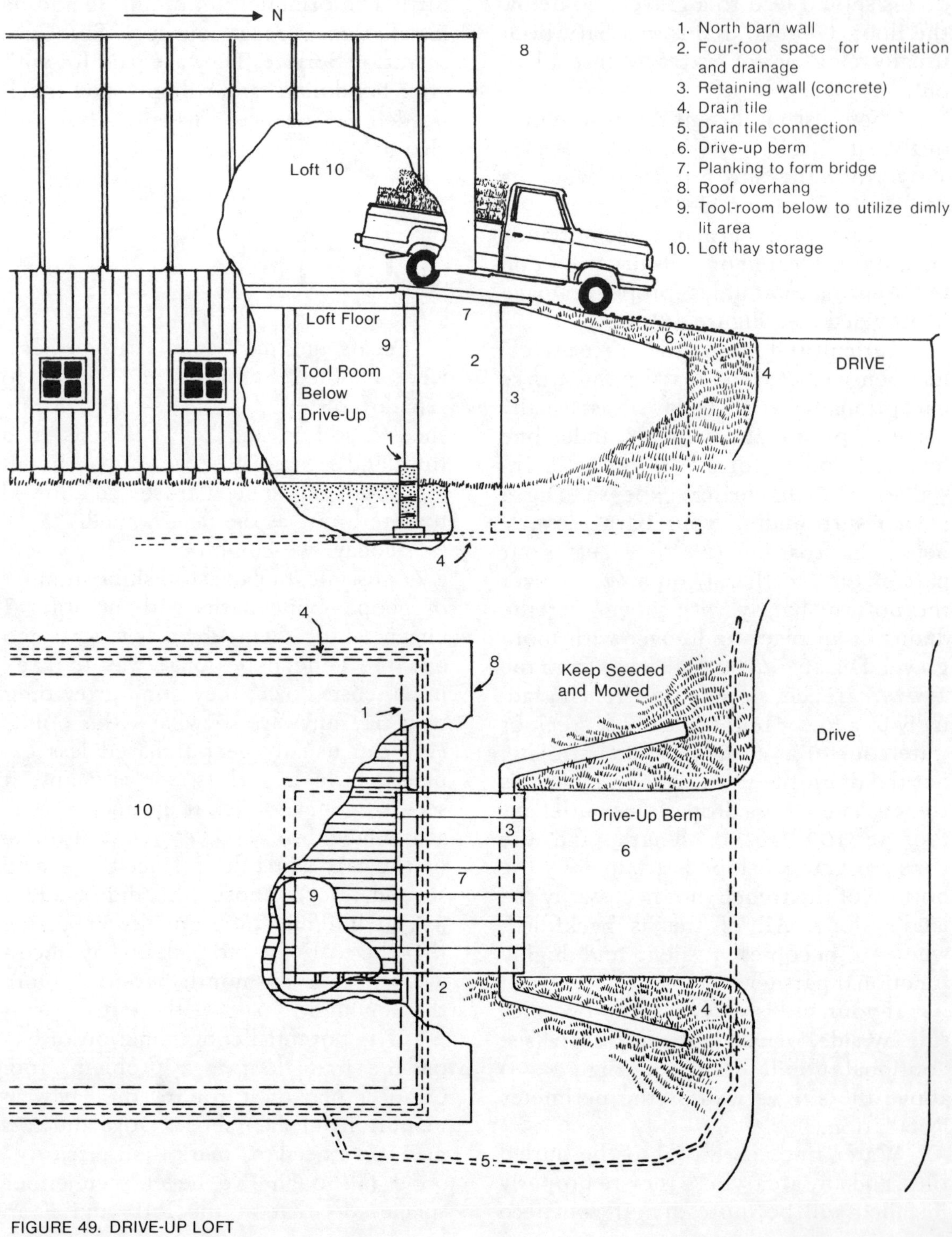

FIGURE 49. DRIVE-UP LOFT

animal gets loose, your weekly hay pile can be ruined in a matter of minutes. If you keep your hay on the first floor and have noticed your stock being picky or downright refusing to eat the hay, these may be some of the reasons.

The taxes are not appreciably higher with or without a loft, and a loft has tremendous advantages as an integral part of the barn. In winter when the loft is full, it retains heat; in summer when the loft is empty, it increases air circulation. As a matter of fact, you can use it as an air conditioning system in summer by leaving the hatches to the loft open. The warm air rushing through the loft will suck in cooler air to the animals below.

No moisture barrier should be put between the animals and the hay in the loft. It will cause condensation on the ceilings and damp barns. Remember the heat exchange upward is greatly slowed by even a single layer of hay bales on the loft floor; thus the hay prevents cold loft air from "falling" to the animals below.

A drive-up loft is convenient and especially easy to build if the barn is located where the ground itself is already on two levels. Plan it so that the retaining wall will not interfere with the barn's natural lighting; located on the north side, it will also act as a windbreak.

To build a drive-up loft, construct a retaining wall two to four feet away from the side of the barn. Run "anchors" on reinforcing rods from high on the wall to the base of the artificial hill. Instead of "anchors" you can extend the concrete retaining wall in a horseshoe shape to retain the soil backfill (see Figure 49); this holds the soil in place even better. (The concrete should be reinforced with wire.) Leave space between the wall and

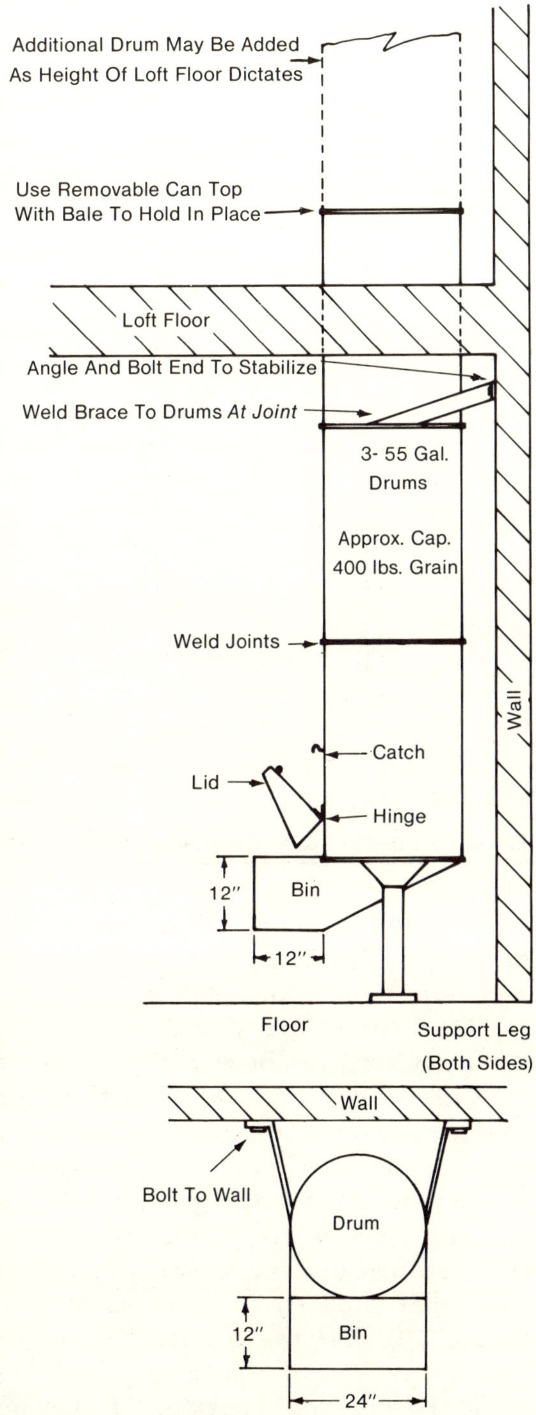

FIGURE 50. LARGER SCALE GRAIN STORAGE SYSTEM

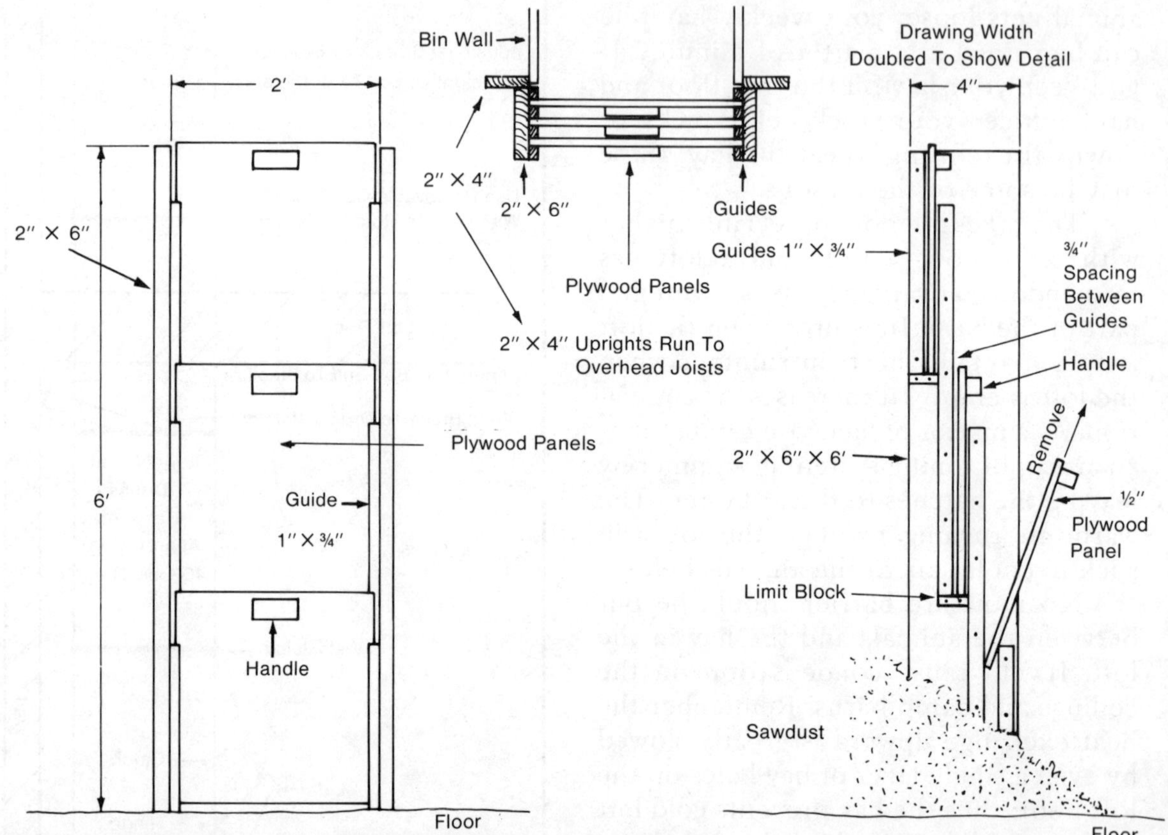

FIGURE 51. LARGER SCALE BEDDING STORAGE SYSTEM

the barn for air circulation to protect the wood structure. Drain tile installation is very important in such a design.

Once items are stored in the loft, gravity systems can be worked out to suit your needs. Hatches in the loft floor for hay feeding can be either centrally located or even directly over mangers. (Remember where they are when storing hay and leave paths through to them.) Grain storage on a bigger scale is handled somewhat similarly in dairy barns (see Figure 50). Use the gravity system for lime or sawdust if that's your bedding choice (see Figure 51). All of this may be too extensive if you only have a few head of stock. But if your few head of stock are growing in number, it's worth considering for expansion.

Don't close loft end walls up tightly. If you use unseasoned boards, nail them loosely together and let them separate by drying. Tightly closed lofts limit free air exchange with stored hay, which is important to prevent spontaneous combustion. You'll get sun-scald on a few bale corners and maybe a little snow. But that air exchange is important, even with dry hay. If you have to put even slightly wet hay in your loft, salt it first. Salt can be purchased in large sacks for a couple dollars at most feed stores. Put it on by hand on each layer of bales put in the loft. Be generous. The salt is cheap and replacing a barn or a full load of hay won't be. Animals prefer salted hay, although most of it falls off. And avoid putting damp hay in the loft in the first place. Damp hay may cause spontaneous combustion.

Roof Designs

There's as much variety in roof shapes as there is in general building. The gambrel roof offers the most room for the least cost in colder climates. It also provides maximum temperature control; because the distance from the roof's surface to the loft floor is so great, you have more control over the inner barn temperature.

In warm climates, high-roofed, open-topped buildings are best for maximum air circulation and to keep animals as far as possible from the sun. The height adds coolness and deepens shade.

Eaves are important and aren't often considered seriously enough by the novice builder. Eaves serve so many functions: they keep water away from the sides of the building, they protect the connection between the first and second floors of a barn, and they protect windows and doors from excessive weather exposure. If too wide, they limit the amount of sunlight available to the animals. If too narrow, windows, doors, and the siding weather very quickly. The size of eaves is directly related to the type of construction you are planning and to the building's height.

Roofing Materials

Each type of roofing has benefits and problems. The list below should help you select which is right for you.

Asphalt shingles make an excellent weather seal, installed over felt building paper. They have a lifespan of twenty to forty years, depending on sun and wind factors, and are moderately priced.

Cedar shakes last even longer than asphalt shingles. However, they tend to rot in a spotty way and are quite difficult to replace in small sections. The initial cost is very high.

Aluminum roofing can last longer than any other type of roofing, but today manufacturers are making it thinner than it should be. Aluminum roofing protects itself by oxidation. This oxidized surface will pit in a hailstorm and the pits reoxidize themselves. A few hailstorms may produce holes in its surface. If the sheets come loose anywhere, nail holes will enlarge because the material is so soft.

Steel roofing can last as long as aluminum if it is galvanized or painted. Acid rain may eventually eat away the galvanizing and you may eventually have to paint the roof or your barn will look like a large rusty can in short order. Ask whether the material is galvanized; the seller may say, "Of course!" but if the price is very good, it may not be. Comparative shop and be certain of what you are buying.

ROLL ROOFING is the least expensive way to roof and must be installed over felt building paper. Carefully seal overlapping edges if you expect the roofing to last. Its major drawback is an "unforgiving" reaction to temperature changes. It shrinks in cold and sags in heat, both of which eventually cause it to crack.

INDOOR/ OUTDOOR SPACE

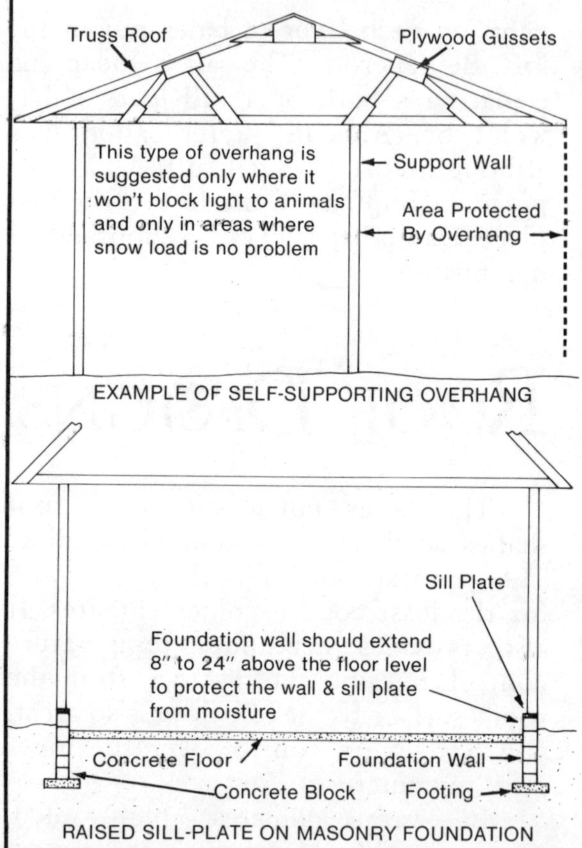

FIGURE 52.

Overhangs on the least-used side of your barn are a valuable addition of space. Used as carports, they help block the wind. Or for animals they may be used in summer for shade and breeze outdoors, and in winter as a place to get out of the elements but remain outside. Overhangs increase your space at minimal cost.

Plan your overhangs just as carefully as you plan the rest of the barn. Their location shouldn't block light into the barn proper by shading windows. If you intend to clean your barn from the outside, build the overhangs tall and wide enough for your vehicle to get in and out. Located in a heavy-use area, the overhang needs a concrete pad—it's worth it. Watch for weather deterioration along the upper edge where the overhang connects with the barn. The barn's eaves will provide some protection, but to lengthen the life and soundness of the overhang, you may want to flash the joint.

Put a good footing under the supporting posts. Overhangs are subject to frost heave because they cover an open space. Unless the posts are properly supported, frost may heave the overhang away from the barn. Don't fasten the posts in concrete. Instead, bolt them to concrete pillars so they can be replaced easily. Fasten them down like you mean

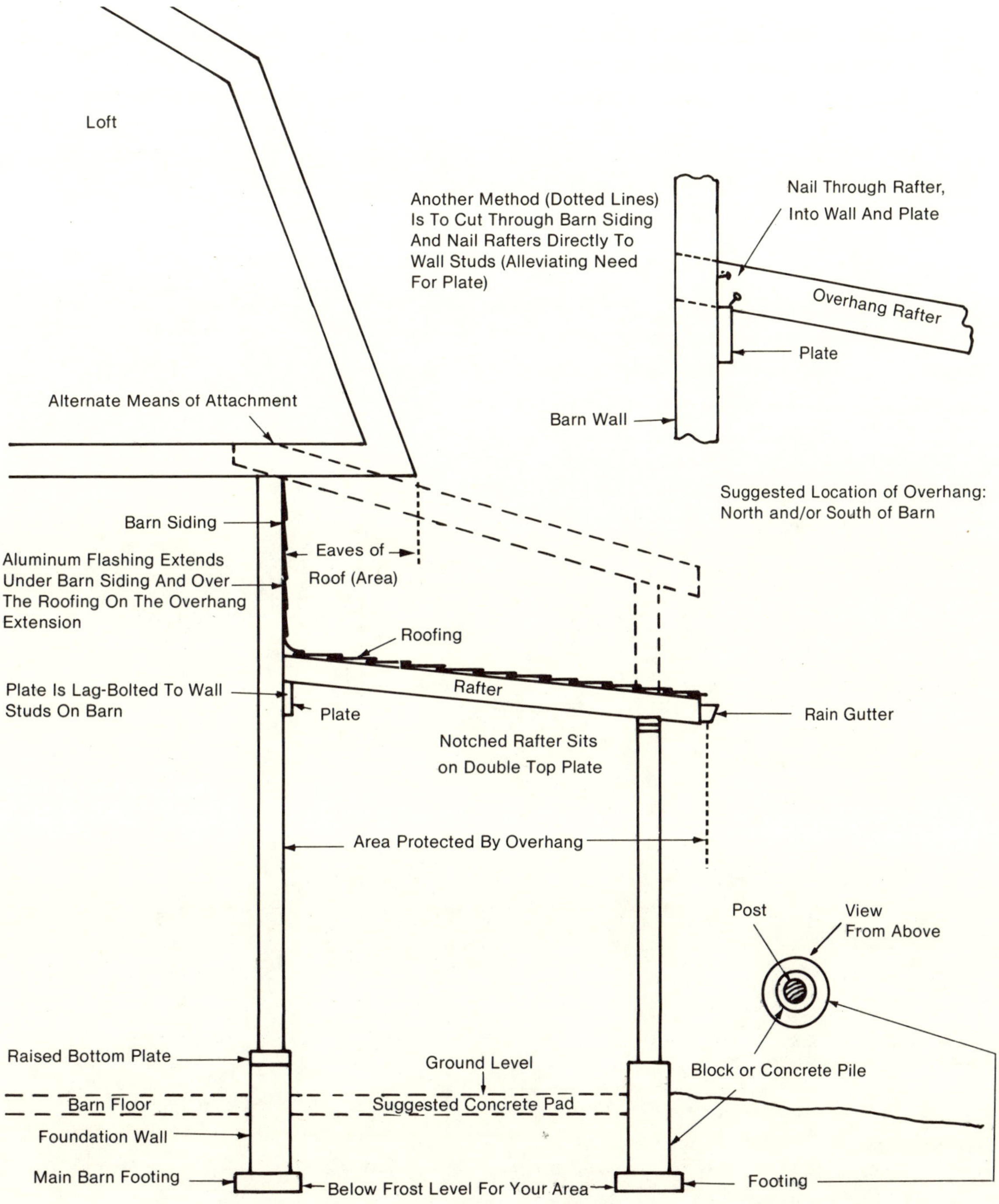

FIGURE 53. OVERHANG ON BARN

it; animals bumping and rubbing on them shouldn't make them loosen suddenly and weaken the whole structure.

If you lack confidence in building your own barn, remember: buildings have been designed and built *by* people *for* people for a long time. And because you're building for yourself, you have a good chance of building better than the contractor who may cut corners.

How To Design It

By now you know what species you want to raise and how to go about raising them. You have a pretty good idea of how to build the barn and install proper drainage around it. You have even decided what materials you want to use. What next?

This is the time to sit down and have a heart to heart talk with yourself and/or with your partner. What is your goal? Where and how far you want to go with what you want to do? For example, you've decided you want goats. Can you see yourself owning two? Yes. Four? Yes. Six? Yes. Eight? Well, maybe, including the kid crop. Got enough outdoor room to do it right? Do you plan to space the does' breedings for a wider spread of milk production or do you want to have all of them deliver at one time to fatten the buck kids for Easter sale? How much room will you need for the kids with the plan you choose?

How about pigs? Do you want to put energy into raising them to the boar-and-sow level? How much will that interfere with your goat program? Or do you just want a little pork? Since pigs raise better in pairs, figure space for two and raise the extra for a friend with a freezer.

How about sheep? Have you always wanted to try your hand at spinning? It will probably take only one sheep to supply enough wool for your spinning try. But sheep are herding animals and a lone sheep in a goat herd will be miserable. So plan for a pair. Have you got enough outdoor space for them? And remember how cute those lambs were that you saw at your friends'? Do you want to expand a little and sell lambs on the side? What about that big lawn—want it mowed? How about four ewes?

So now we've taken care of your pork, lamb, and milk needs. What about hamburger, steak, and roast beef? Besides, nothing impresses your city relatives as much as the size of a cow.

Maybe a steer. And only one because they eat so much. Pasturage makes a vast difference in the price of your homegrown beef, so do you have enough room for him?

The children have always wanted a pony, and it might help out in a pinch to have an animal that can move small amounts of firewood, right? Will your building's eaves be wide enough to store your pony cart under? Or do you need to plan room inside the barn?

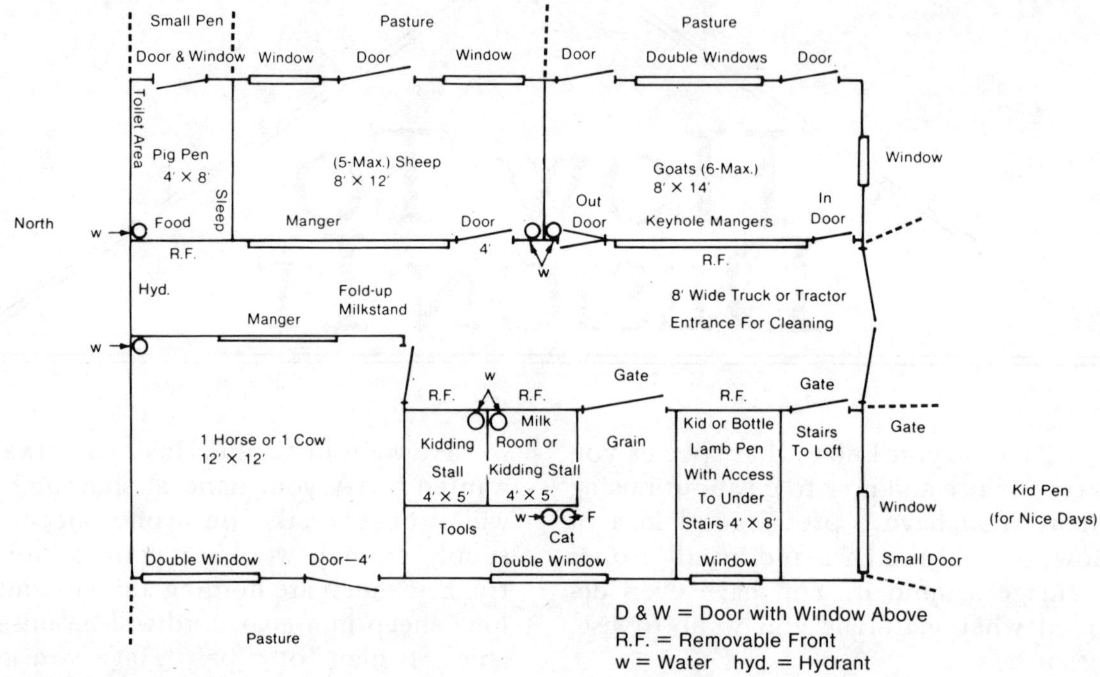

FIGURE 54. 24′ X 30′ BARN

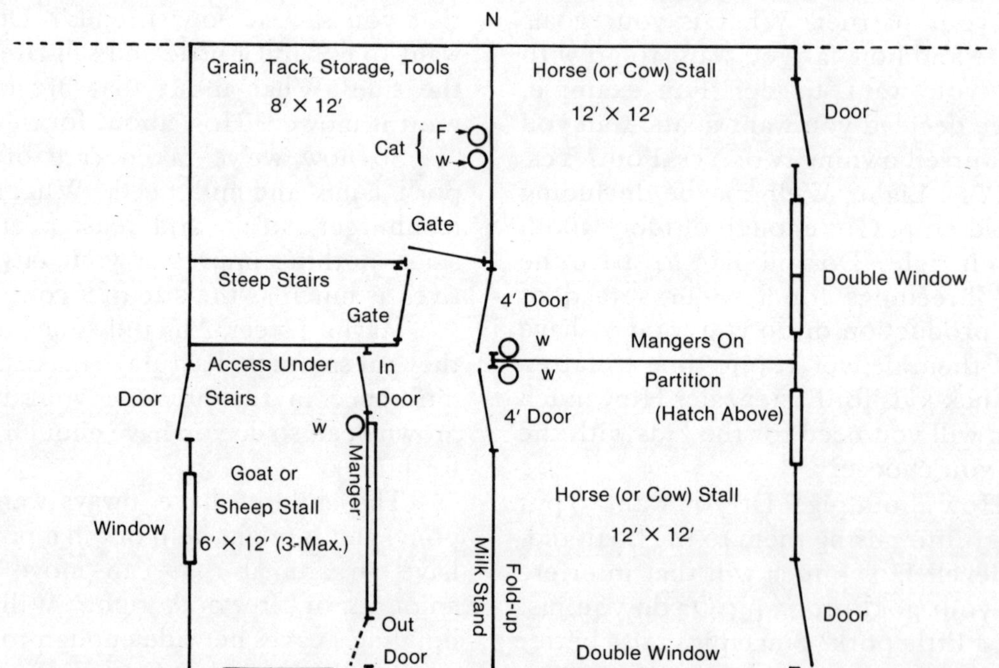

FIGURE 55. 24′ X 24′ BARN

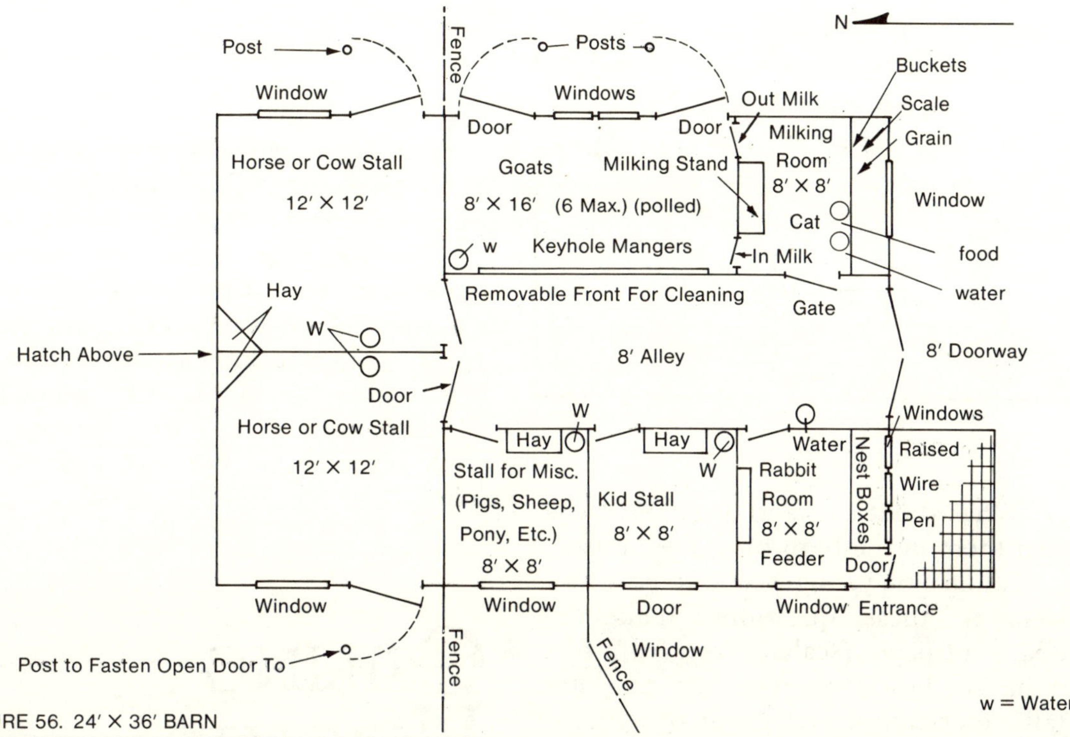

FIGURE 56. 24′ × 36′ BARN

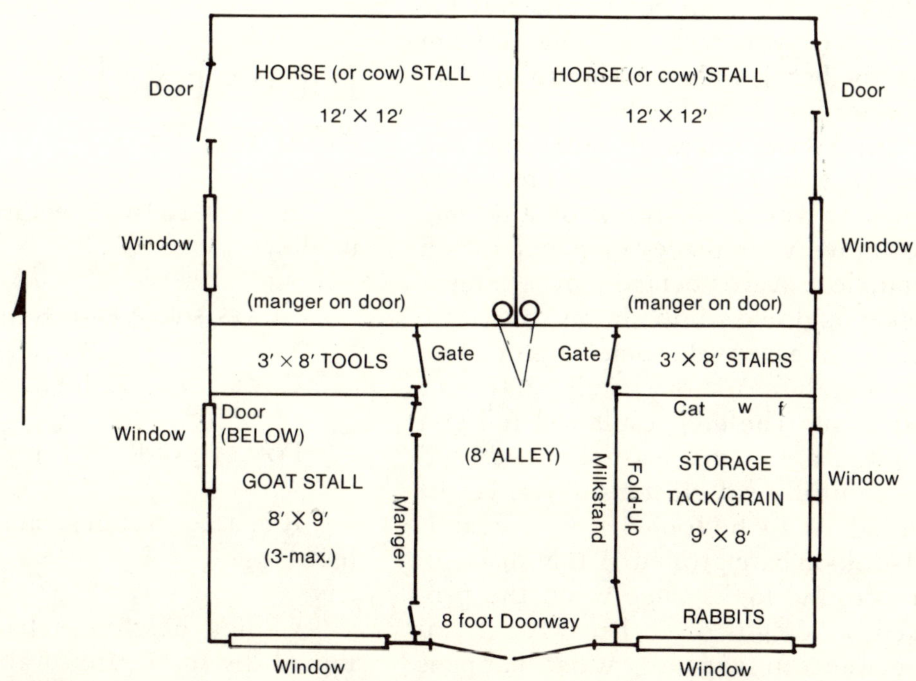

FIGURE 57. 24′ × 24′ BARN

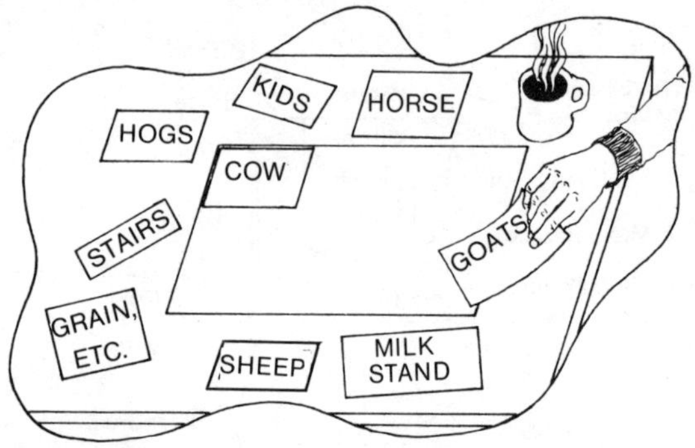

FIGURE 58.

What about cats and dogs? Will you keep them in the barn or in the house?

When you've figured out your answers to these questions, make up squares of paper (scaled down from feet to half-inches) that resemble the size stalls you want for each group of animals. Add to these, room for feeds or tack, milkstands, room to get in and out for cleaning (if you will be doing it from *inside* the barn), and any other things you feel are needed. Make those into representative scale pieces of paper and add them to your pile. When you're pretty darn sure you can't think of anything else, move your pieces of paper into a rectangle, square, octagon or whatever shape you'd like to do chores in twice a day. You'll have to do some juggling, like changing the shape of a "stall" to fit into your plans. The end result will tell you *how big*. Take a tape measure outside and figure out if it will fit the site you've got in mind. *Do not* be fooled at how small it will look on bare ground. A tiny space on bare ground looks huge when the proposed building is on it. But what if you ever want to expand? What happens then?

Partitions can change location and style. Barns can expand most easily if they are rectangular, since you can usually keep your width intact (narrower dimension) and expand your length without interfering with your basic layout.

For example, a 24 × 30 barn can become a 24 × 60 barn with little or no alteration in your initial construction. If you've got a sneaking suspicion that you just might expand someday (animals of all kinds have a way of "growing" on you), select your barn site with this in mind.

Ceiling Heights

Ideals for Each Species:

HORSES: 10 to 12 feet in stalls, 14 feet in alleys.

GOATS, SHEEP, PIGS: 8 feet.

COWS: 10 to 12 feet.

PONIES, DONKEYS: 8 to 10 feet.

RABBITS: (in stall arrangement) 6 to 8 feet.

Ceiling heights in barns are determined by moisture evaporation, heat retention, the species' habits, and specialized problems. *The majority must rule.* That is, the single horse in a barn full of goats and sheep should have an eight-foot ceiling. To make this combination more compatible, locate a loft entrance above the horse's stall which can be opened for air ventilation when necessary (see Figure 65).

Should the reverse be true (two goats and ten horses, for example), a false ceiling above the smaller animals may be necessary. Building a hay storage rack seven or eight feet above the goat stall, for example, provides them with easier heat retention and yet keeps the ceiling high enough for the larger animals. Adjusting the height of the loft floor in various areas of the barn has many problems attached to it: it reduces the flexibility of space in the barn, and the necessary carpentry skills takes it out of the "build it yourself" category. It also diminishes the structural strength of the barn no matter who does the work. And it's unnecessary. Again, we're discussing *ideals*, not laws.

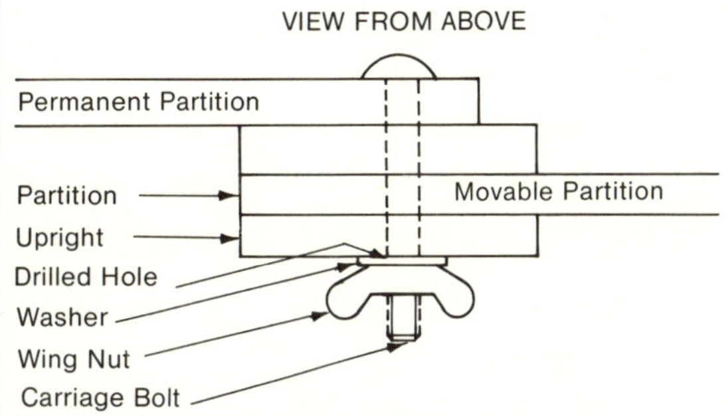

FIGURE 59. ATTACHING MOVABLE PARTITION

Partitions

Partition heights vary from species to species for good reason: some animals would never dream of leaping a partition; others dream of nothing else. In each stall at least one internal wall should be solid to floor level. When it's cold, animals will choose this wall to sleep against because it has no draft underneath and it isn't as cold as the outside walls. This is most often the wall that separates the animals from the barn proper, instead of the partition between animals.

Support board partitions between stalls top and bottom with "two-by" lumber (2 × 4, 2 × 6, 2 × 8, etc.). Then nail partition boards to these supports. Vertical boards wear much better because if an animal works on a single section of wall, the most that can happen is a single board will become loose or fall off. When the boards are horizontal, the partition may be lowered or raised enough so the animals can leave the stall. Further, vertical boards shed water better and dry more quickly when wet by the wicking action of the wood grain.

Partition Heights: Ideals

Horses: 5 to 6 feet (topped with bars or wire).

Ponies and donkeys: 4 to 5 feet.

Pigs and hogs: 3½ to 4 feet.

GOATS: 4½ to 5½ feet.

KIDS: 4½ to 6 feet. (Variations depend on stall items that may be climbed.)

SHEEP: 3½ to 4 feet. (A few breeds can jump well.)

COWS: 3½ to 4½ feet.

SQUARE FOOTAGE PER ANIMAL

I strongly advocate leaving animals outdoors as much as possible all year long. Locating water in one place, salt in another, and so on, will encourage all animals to get outdoors regularly.

Because stalls are normally designed by a single given measurement (such as a basic 10 feet or a basic 12 feet*) I offer the figures below as examples. Then make whatever conversions are necessary for the barn you are designing.

DOORS AND SAFETY

As a general rule, all *in*ternal barn doors should open *in* to the animals; all *ex*ternal doors should open *out*, away from the animals. By internal, I mean those leading into the stall and other closed areas. External doors are those leading from the barn to the outside.

STALL SIZES
(Minimum Space Required Per Head).

Species	*Minimum Square Feet*	*12′ Conversions (in feet)*	*10′ Conversions (in feet)*
Beef cattle***	60 sq. ft.	2 animals/10 × 12	1 animal/6 × 10
Dairy cow***	100 sq. ft.	1 animal/8 × 12	1 animal/10 × 10
Dairy goat***	20 sq. ft.	6 animals/10 × 12	5 animals/10 × 10
Hogs	40 sq. ft.	4 animals/14 × 12	2 animals/8 × 10
Horses	144 sq. ft.	1 animal/12 × 12	1 animal/14 × 10
Ponies or donkeys**	60 sq. ft.	2 animals/10 × 12	1 animal/6 × 10
Rabbits (breeders)	8 sq. ft.	6 animals/8 × 6	10 animals/8 × 10
Sheep	20 sq. ft.	6 animals/10 × 12	5 animals/10 × 10

*Or multiples thereof. The stalls must fit within these basic limitations.

**Anytime two of this species are housed together, make certain they are friendly with each other first. (See Species Specifics.)

***Multiply minimum floor space by three for horned animals. Or house each horned individual animal separately. Or dehorn them.

INTERNAL DOORS

You have the maximum control where doors open *in* to the animals. Animals soon learn that crowding the door doesn't mean they can get out more quickly; in fact, they must step back or be pushed back by the door. The door itself becomes the controlling buffer. And when you aren't present, doors which open *in* prevent animals from just bulling them open through sheer power. Instead of the latch alone holding the animals in—as in the case with a door that opens away from the stock—the whole door is forced shut when they push on it. When entering the stall yourself, the animals are forced to step back. It's easy to reach back over and latch the door shut; the animals cannot crowd past you.

The only potential problem with internal doors that open *in* to the animals is manure or bedding build-up which blocks the door. This is easily solved by elevating doors six to eight inches above the floor level. (To reduce the possible drafts with this arrangement see Figure 60).

Location of an internal door also affects the control you have of that door. Where possible, locate the opening edge of the door in a corner. This allows only a single animal to get close enough to try to get out. A corner location also makes it hard for them to work on the door when you aren't there. As you enter and leave the stall, the door forces collected animals off to the side and out of your way. (If a sheep or goat is in your way, tip her head back with your fingertips under her chin.)

The size of the door is important. One tends to think of household doors as "door size" without ever thinking about the barn door's intended use. A smaller door is stronger, easier to control, and a lot easier to keep closed; the hinges can be lighter too as the weight of the door is less. If smaller animals almost exclusively use the door, a foot and a half wide is all that's necessary.

Smaller doors are advantageous with milking goats. Separate the two doors in the goat stall as far as space permits; then let your milkers out one door and return them through the other. In a short time, the goats won't even try to gain passage through the "return" door, and you will never again have to stuff a doe into a group that's trying to get out for their turn at milking.

Smaller internal doors can be of benefit in still another way. They allow the smaller, younger animals to escape the pressures of the larger ones for resting, eating, and safety—and this helps them to grow. (The same principle is used in creep feeders, where only the young can fit through the entrance.) I had such an arrangement once for goat kids before learning the benefits of separating the goat kids completely from the adults at all times. A narrow baffle doorway, which provided a six-inch opening, could be adjusted down to three inches for the kids when they were

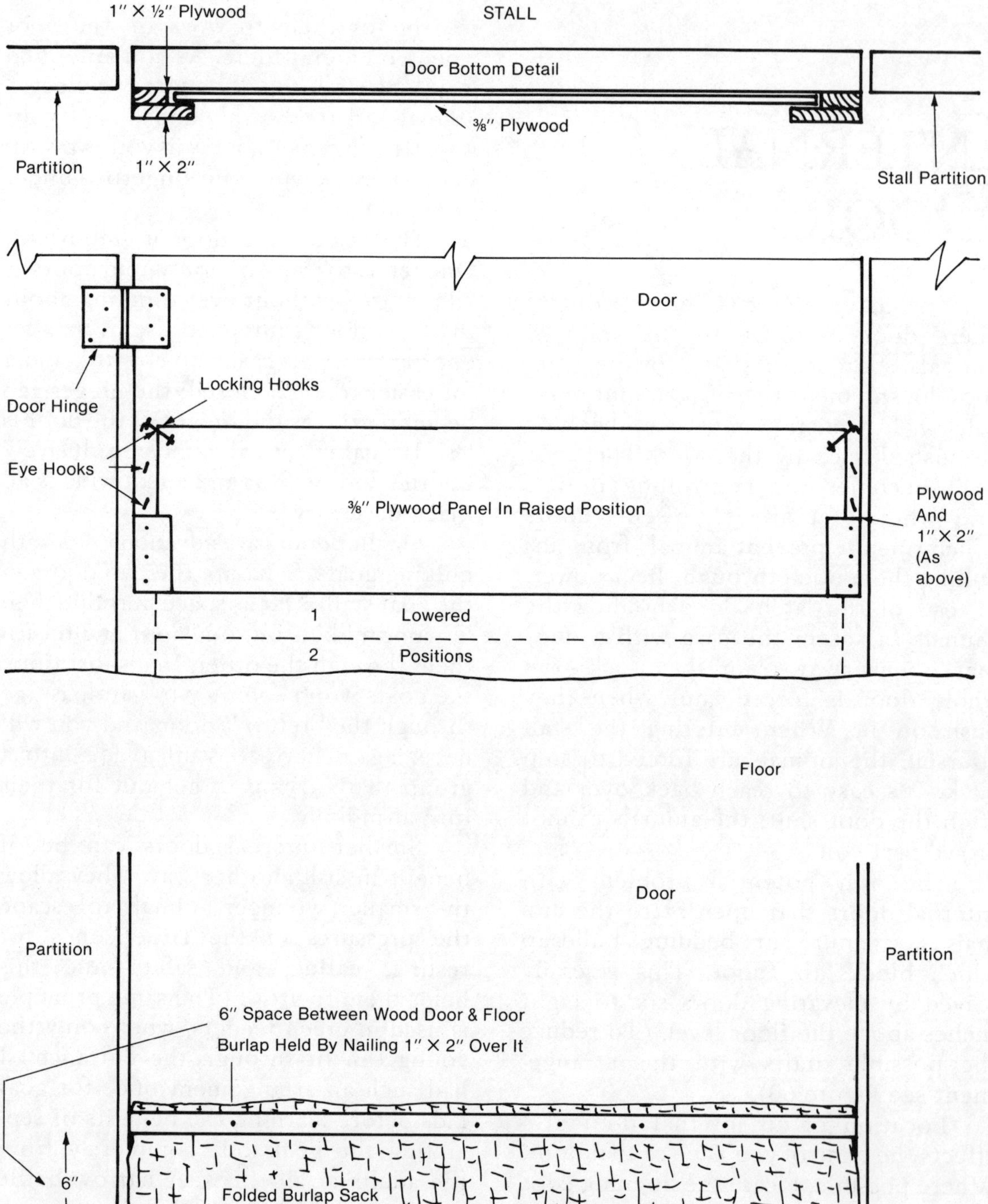

FIGURE 60. DRAFT-FREE STALL DOORS

quite small. (See Figure 61 for details.) An opening any larger than six inches and, believe it or not, the does could force their way in! At all times, the stall contained hay, grain, and fresh water. It was ideal for sunny, communal naps, and the kids used it often.

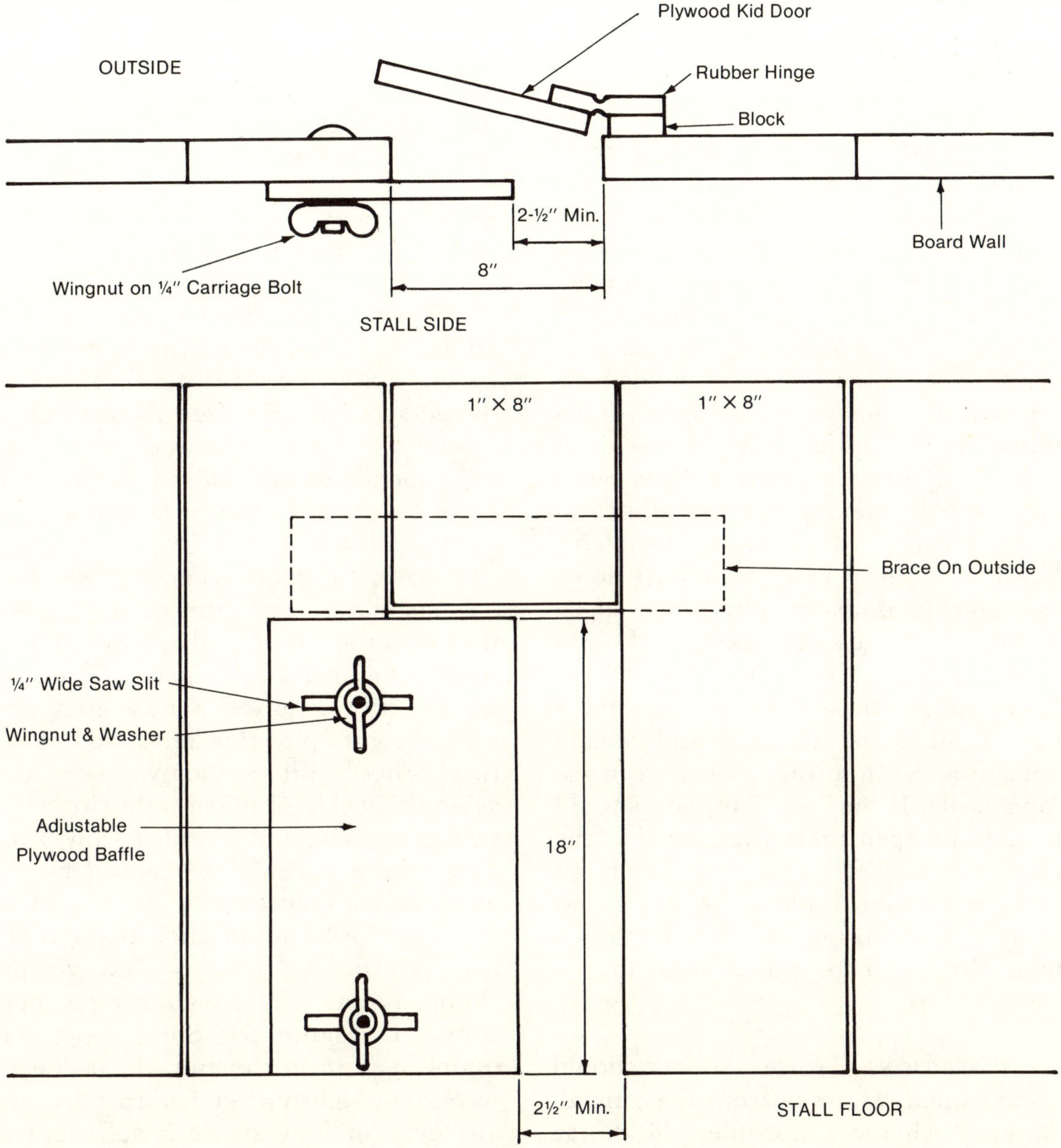

FIGURE 61. ADJUSTABLE KID DOOR

External Doors

External doors, as stated earlier, should as a general rule open *out*, away from the animals. The reasons are exact opposites in many ways as those of internal doors. Since the animals can and do crowd openings, there is little or no chance of an animal becoming pinched in a door they are trying to leave the barn through. For this reason, external doors should be a minimum of four feet wide in most cases. Bossy animals can less easily block a wide opening, keeping all others in or out. (If you are having problems like this, either break the group down into smaller units to relieve the social pressure or put a second external door in farther down the wall.) Another important reason for these doors to open out is fire: if the doors are wide and remain open for free housing, animals can escape easily. If they are shut, you should be able to open them from outside the barn, which will keep you from having to enter a burning building. There is no greater terror to the stock owner than a barn fire, so take precautions in advance.

Exceptions: Horse doors should always open *out*, away from the animal. In fact, wherever possible, the horse should have a straight path to the outdoors, without small corners and sharp turns. Horse doors should swing out and against the wall where possible.

The other exception is kidding or lambing stalls. A doe or ewe may lie directly against the door when birthing begins. You will have to get in to assist her or to remove the young, and you don't want to disturb her. It's a real help under such circumstances when the door opens out and you can slip right in.

Door Latches

We use three types of door latches. On larger doors, such as on horse stalls, we use a two by four slotted lock with a thread-spool handle (see Figure 62). It should slide twelve inches or more from fully opened to fully closed. In that way, a horse knocking a door constantly can not open it eventually. Or you can slip a chain over the spool when it's closed.

Years ago, we came upon a hook that works so successfully, I would have sworn no animal could open it. However, we recently acquired a new goat who walked right up to this latch and opened it. This hook still has plenty of uses such as on the inside of horse stalls (for when you're in there) and outside of any other door where it can't be reached by the animals (see Figure 41).

The third kind of latch that (so far) has been successful for us in every application is the chain-snap arrangement shown in Figure 63. Since most stall doors open in to the animals, it doesn't have to be tightly stretched to work, and an "eye" or long staple is sufficient to hold it closed.

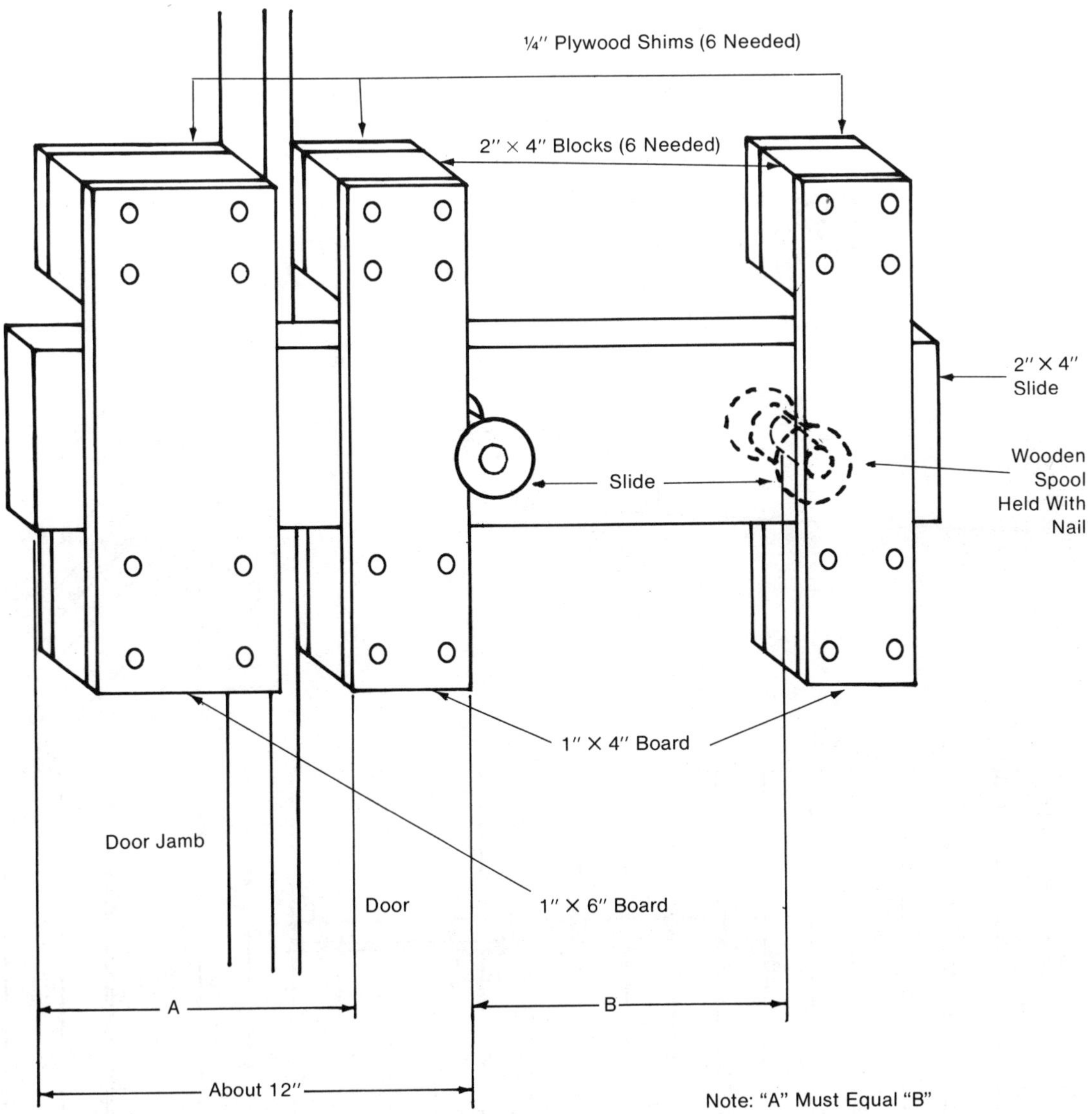

FIGURE 62. SLIDING LATCH FOR HORSE STALL

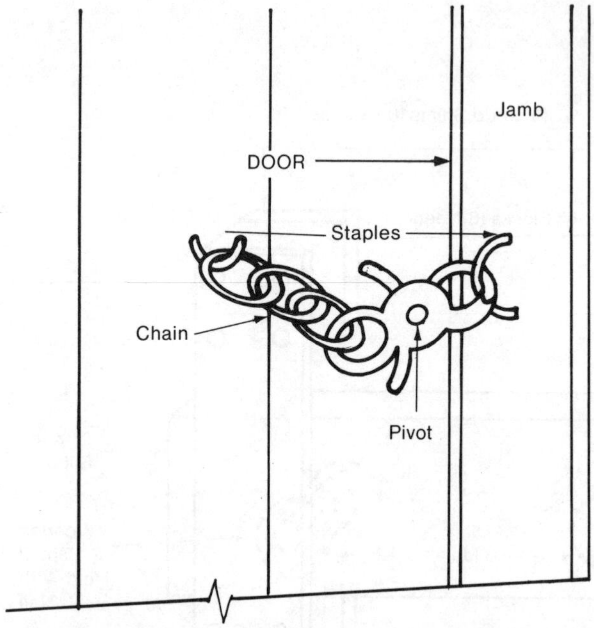

"GOAT PROOF" STALL DOOR LATCH

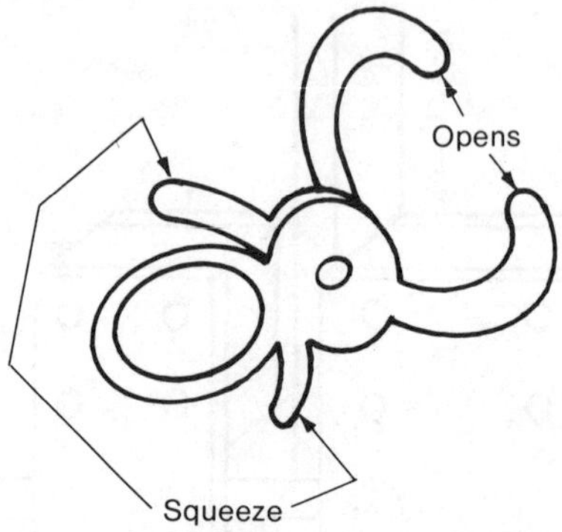

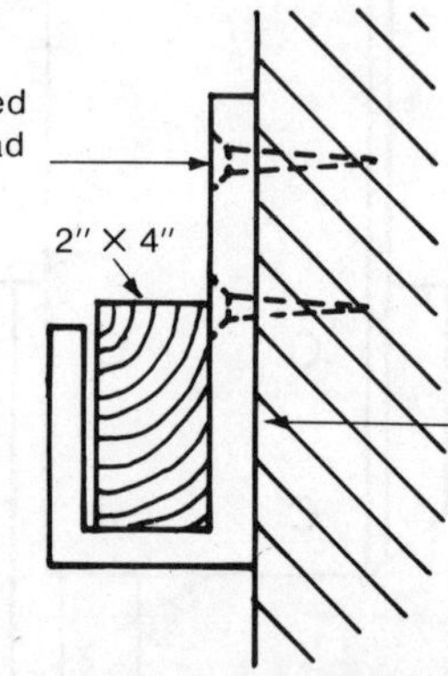

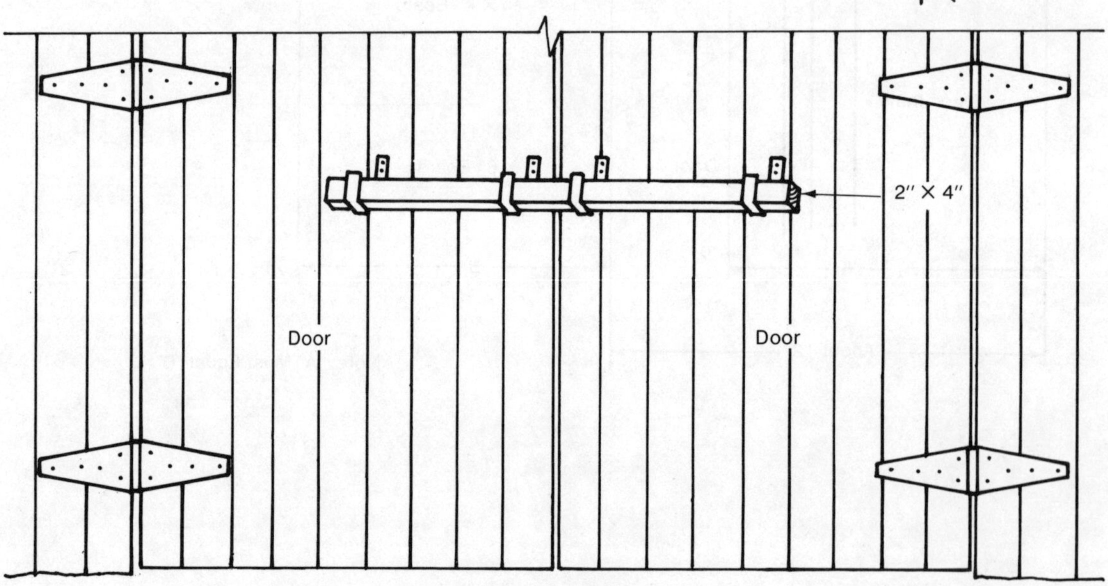

MAIN BARN DOOR LATCH

FIGURE 63.

Stairs and Gates

Stairs in houses are designed to be climbed frequently. They are usually long, over a length of floor space and wide enough so furniture and things can be taken up and down them easily. But this doesn't apply to your barn where there are also loft entrances outside.

Hatchways in the loft floor make "bringing" hay down easier. A single hatch, centrally located, may be sufficient. Many old horse barns permit hay to be dropped from above directly into mangers; this way of haying can be used wherever sufficient amounts of hay are fed at one time. But don't forget to leave paths to these hatches when filling the loft with hay, and build the hatches between beams (see Figure 65).

A stairway is desirable in the barn, if only for going up to drop down a day's hay supply. Design the stairway with this in mind; do not design it so that anything of size may be carried up and down the stairs. Normally a ladder of sorts built right onto the wall is sufficient. A very steep stairway is nicer to climb than a ladder, and it doesn't take up much room.

Keep the stairs narrow and short, the space under them can be utilized as a milking room for goats, tool storage area, or kids' pen.

Internal barn gates provide additional protection. You may plan on no "escapes" but they happen occasionally and having a small light gate across areas designated "for human use only" will keep these incidents from becoming serious occurrences. Such a gate across the bottom of the stairs will discourage that clever goat escapee from contaminating the whole year's hay supply in a rowdy afternoon. The gate should be light enough to interfere little with your trips up and down. Gates which close *to* the stairs make it easy for you to push the gate aside but hard for an animal to open—and perhaps it will delay a loose animal long enough for you to discover her escape!

Small gates mean lighter hinges can be used, since the gates weigh less. This applies to the smaller doors on animal quarters as well. Long ago, we came up with an essentially free hinge which can be made instead of purchased, and will spring the gate or door shut (see Figure 66).

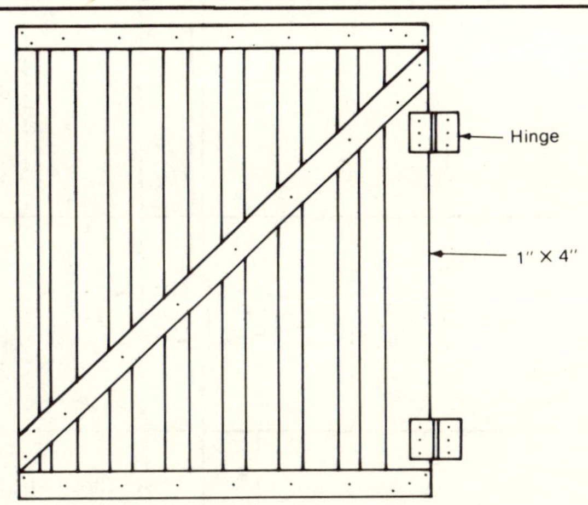

Constructed of 1" X 2"s with the exception of one 1" X 4" to carry the strength to the hinges.

FIGURE 64. LIGHT-WEIGHT GATE

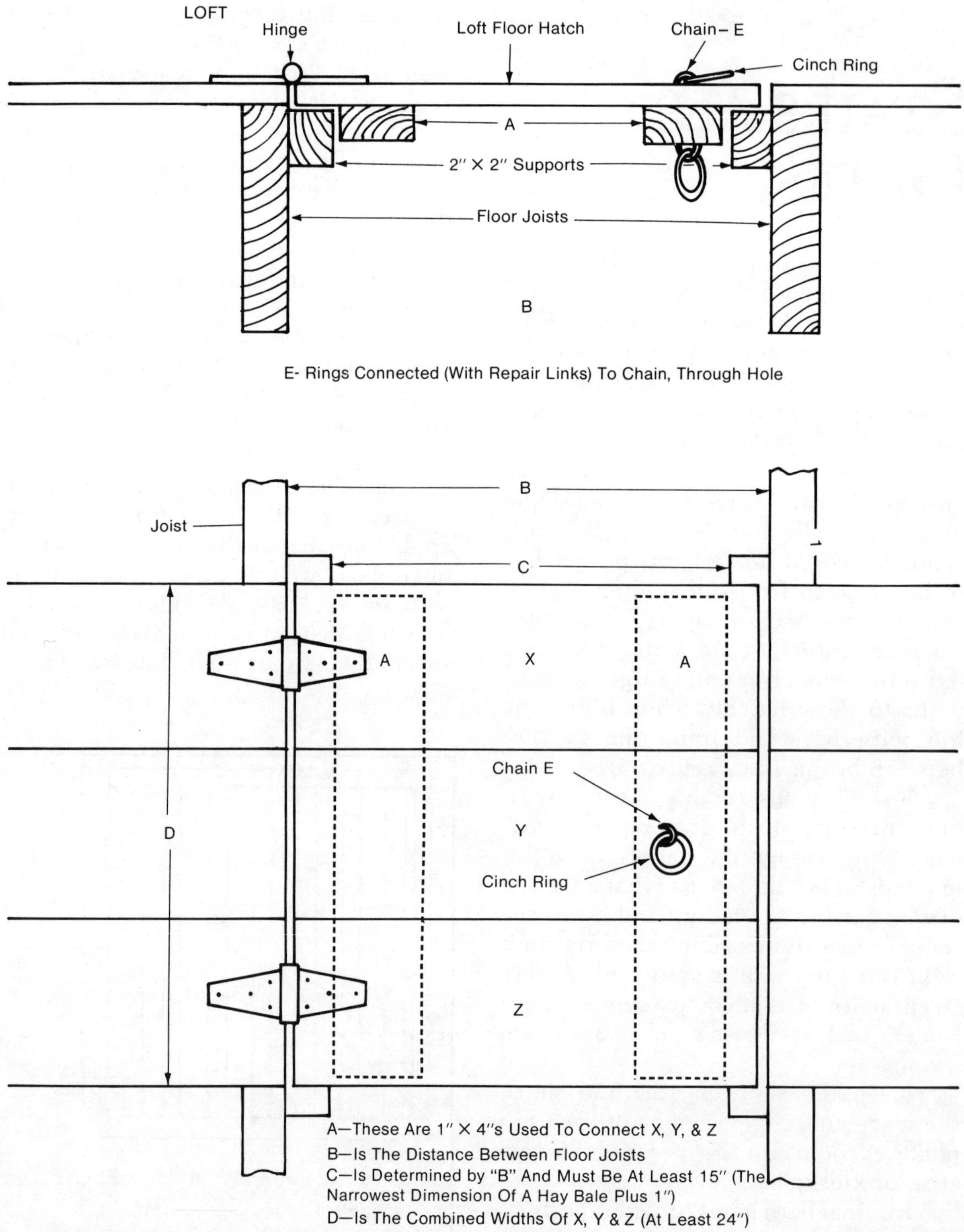

FIGURE 65. HAY LOFT FLOOR HATCH

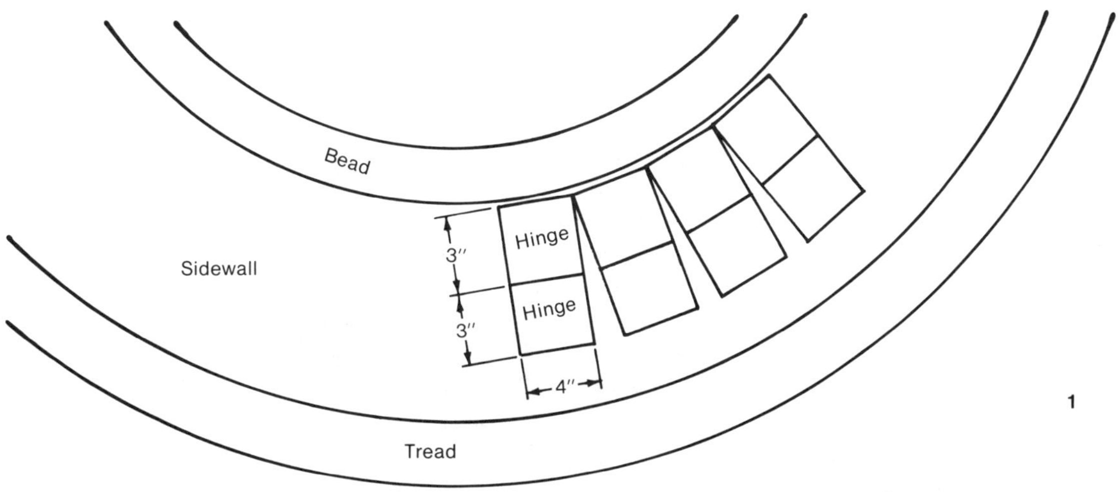

For the hinge you will need an old tire. Drawing *one* shows where to cut hinges; the rubber is stiffer in that portion of the tire. Cut them three to five inches long. Drawing *two* shows two parallel knife slices in the first layer of rubber, about a quarter of an inch apart, running lengthwise down the hinge. *This must be done on both sides*, one above the other for flexibility.

Drawing *three* shows rubber chipped out to the first set of nylon cording *on both sides*, leaving a thinner slot for the hinge to bend on. Nail them to the door with ringshank nails about 1-1/2 inches long, or use screws. The nails make less of a hole in the rubber, and are quicker to use.

These hinges last a long time and used in twos or in threes will hold the weight of any light door. We've had a trio of these hinges on a hay manger gate inside the barn for more than six years; they have only recently failed to spring shut.

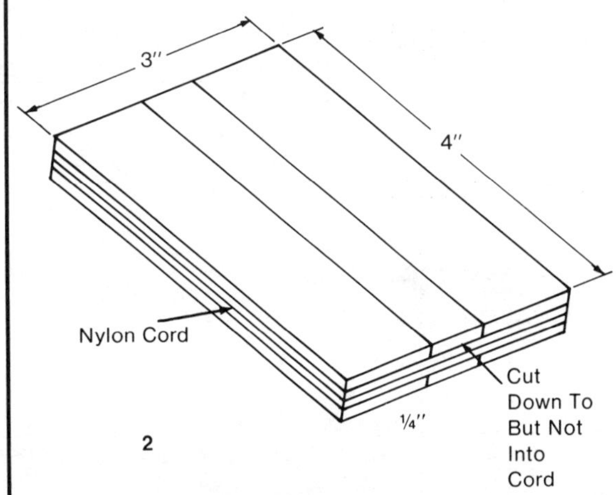

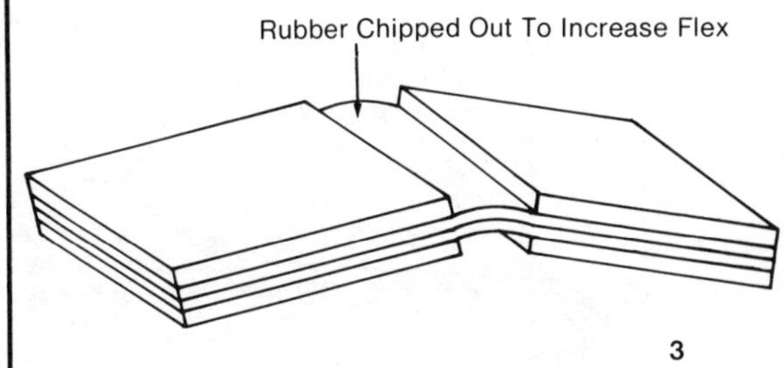

FIGURE 66. HOW TO MAKE RUBBER HINGES

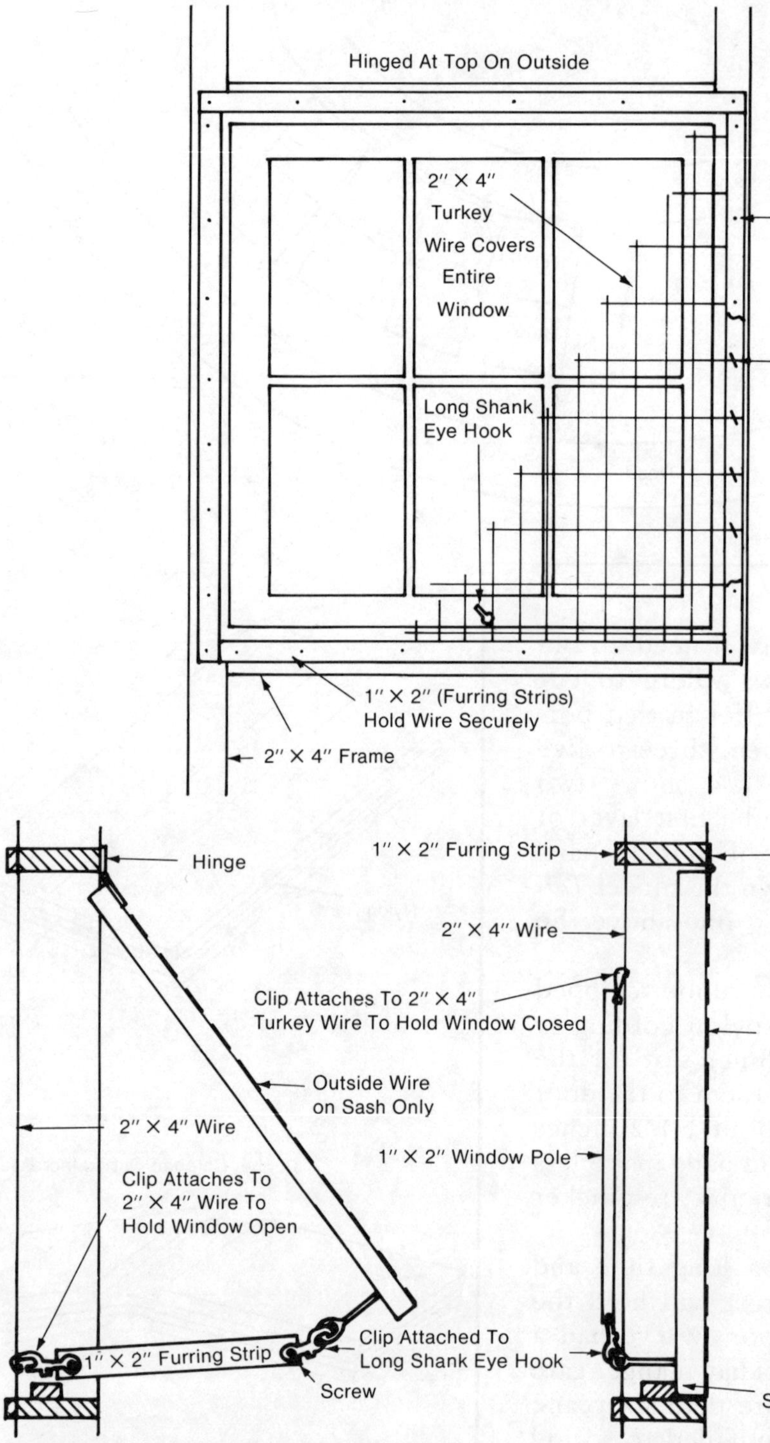

FIGURE 67. PROTECTING WINDOW GLASS WITH WIRE

WINDOWS

The importance of natural light in a barn should be stressed. A window streaming sunlight into animal housing accomplishes so much! It kills bacteria, helps dry dampness, and is vitally important to the normal functioning of an animal's eyes. Besides all that, it creates a feeling of well-being and peace.

Dark barns where animals aren't let out often enough can cause blindness. Retinal function is impaired when animals spend long periods in a darkened barn.

Locate the windows about five feet above the stall floor. Cover them on the inside with wire pulled tightly enough so that the animals can't get ahold of it to pull it loose. Chicken wire works well. Two-by-four welded rectangular wire is also good. Pin down the wire edges so no animal can get snagged or can pull the wire off.

As a rule, animals don't pay much attention to windows from the outside of the barn if they can come and go as they please. However, if you can afford glass windows, cover them on the outside with wire fastened to the frame. This can be done to plastic windows as well. Smaller-paned windows are best because it's less expensive to replace a pane. The smaller panes and wire coverings prevent an animal from sticking its head through broken glass should it break a window.

An excellent source of windows of any kind are auctions. Watch the local paper for such sales prior to barn building; if you have already built your barn, it still pays to buy windows this way, although you may have to modify your openings to fit what's available. Leftover windows make nice coldframes.

Plastic on windows works very well, too, but it's not as permanent nor as transparent as glass. (Some plastics limit what rays of sunshine can come through.) Window openings without windows (but with the frames still covered with wire) can be covered with plastic in the colder months. Water running down the side of the building, however, may collect around the plastic on windows and begin to rot the wood. Fasten down the plastic edges with wood to prevent blowouts in higher winds and, most important of all, to prevent animals from tearing bits of it off and eating it. Plastic clogs the intestines and can cause death. Some plastics are toxic, too.

Double-hung windows don't work very well in barns, at least not in their original arrangement. Cobwebs, hay chaff, and general junk clog the sash channels. Separate them and use them as individual windows. If they're installed to tilt in or out to open, they'll be difficult to protect with wire. And if they are hinged at the bottom to open out and aren't well under the barn eaves, rain will collect on them and pour into the barn. See accompanying sketches for ideas (Figures 67 and 68).

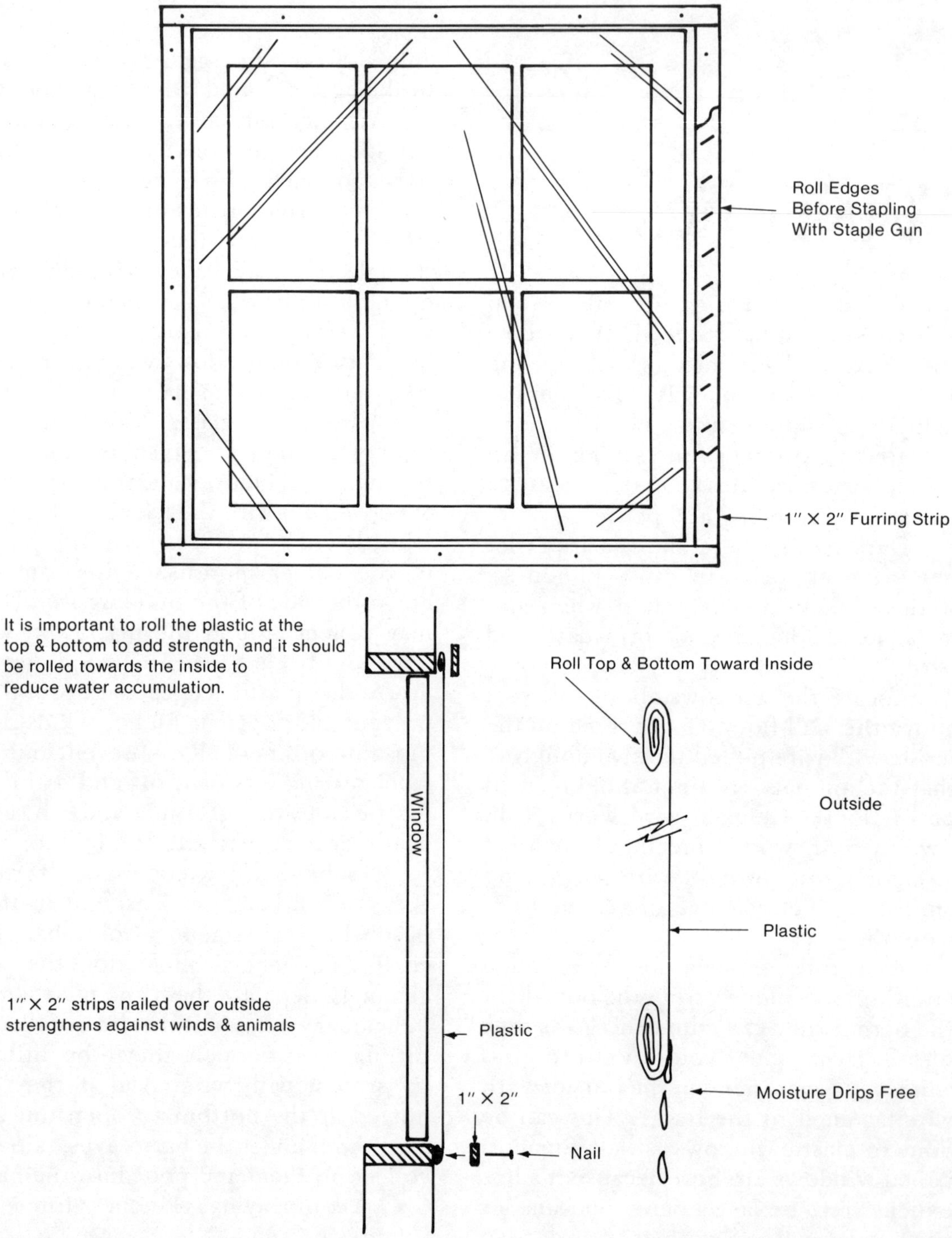

FIGURE 68. WINTERIZING WINDOWS WITH PLASTIC

WARMTH

Those of us not raised with livestock all our lives want the barn to be warm for the animals so they are comfortable—with due consideration for our own comfort when we feed and milk them. When we let the animals out of a warm barn and notice they are cold, we consider this strong evidence that the barn should be still warmer, and proceed to close it up tighter than before.

Actually, if your animals are comfortable in your barn but show signs of stress outdoors on a sunny day with little wind, open a window instead of closing the barn tighter.

Temperature *difference* is the real culprit.

The function of animal housing should be to ease temperature differences—*not* to keep the animals warm. It isn't cold weather that makes animals suffer; it's dramatic changes in the weather or an extreme temperature difference between housing and outdoors—and that goes for cold or heat. In one crazy February here, the temperature went from ten degrees to sixty degrees in twelve hours. That's the easiest weather to lose stock in. The function of the barn in that instance was to ease the change by being somewhat warmer than the outdoors when the temperature was low, and holding the cold longer when the temperature rose suddenly.

It pays to understand how animals keep warm. Most livestock are cold-climate animals to begin with. They grow their "long underwear" not in response to the weather, but in response to *the length of daylight available*. As the days get shorter, their hair grows longer—or woollier or thicker, depending on the species. The process is reversed in spring. This is an automatic process, activated by the pituitary gland, and is essentially the same for all livestock.

The process varies only when the quarters are too warm for too long. Our rule of thumb is a three day limit: any animal who has been in heated quarters for three days or more must be returned to "normal" weather over a slow reintroduction period.

The basic principle you're after in your own barn is to allow the *animals to retain their own heat* as opposed to having a warm place to go. What your animals need most of all in their housing is a windbreak, no drafts, dryness, light, and proper ventilation.

DRAFTS AND VENTILATION

A ventilation system is controllable air movement which is adjustable according to the weather. Ventilation involves an air exchange which keeps air fresh, reduces moisture in air and bedding, and does these things without robbing an animal's body heat.

A draft is an ever-present and un-

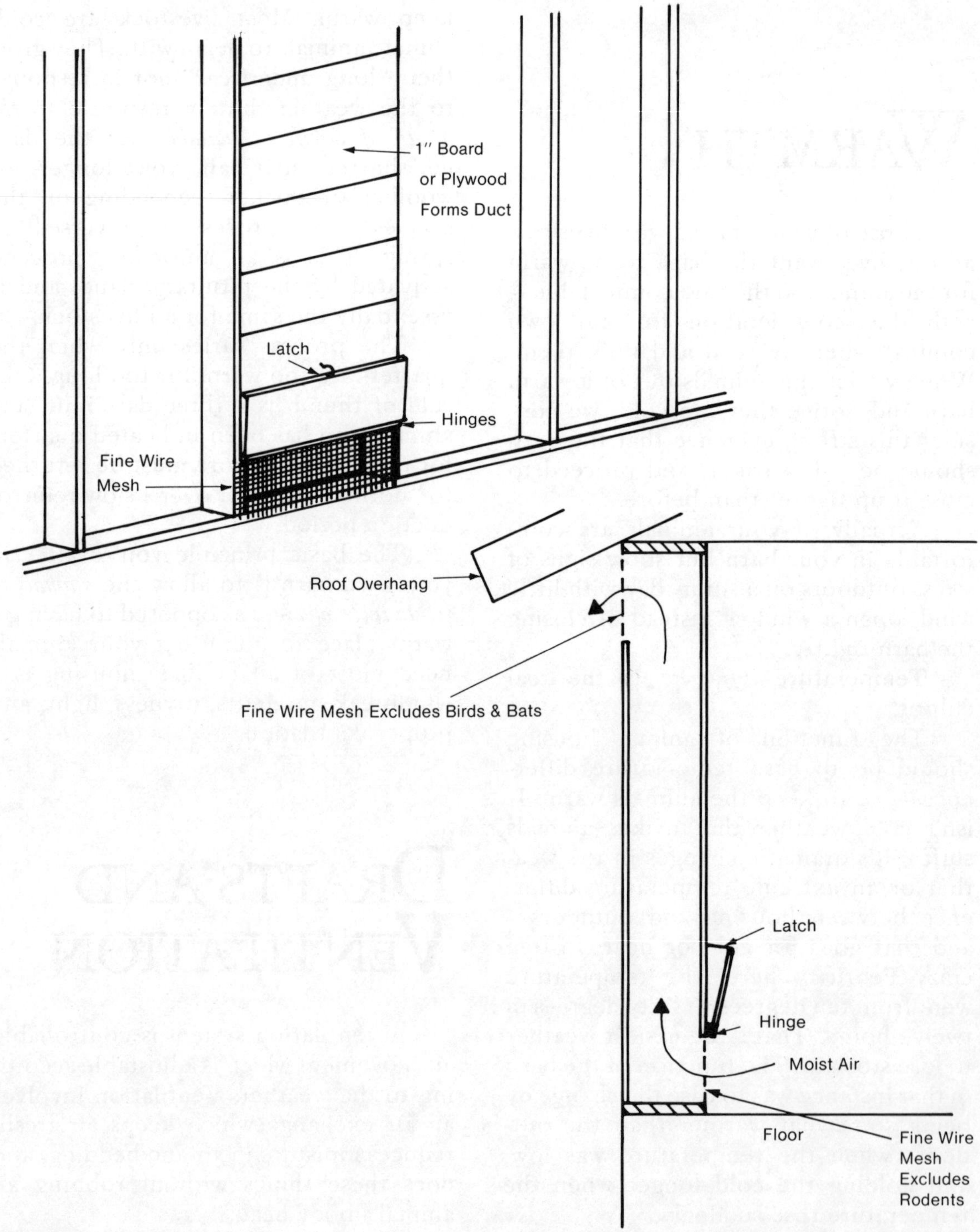

FIGURE 69. WALL VENT

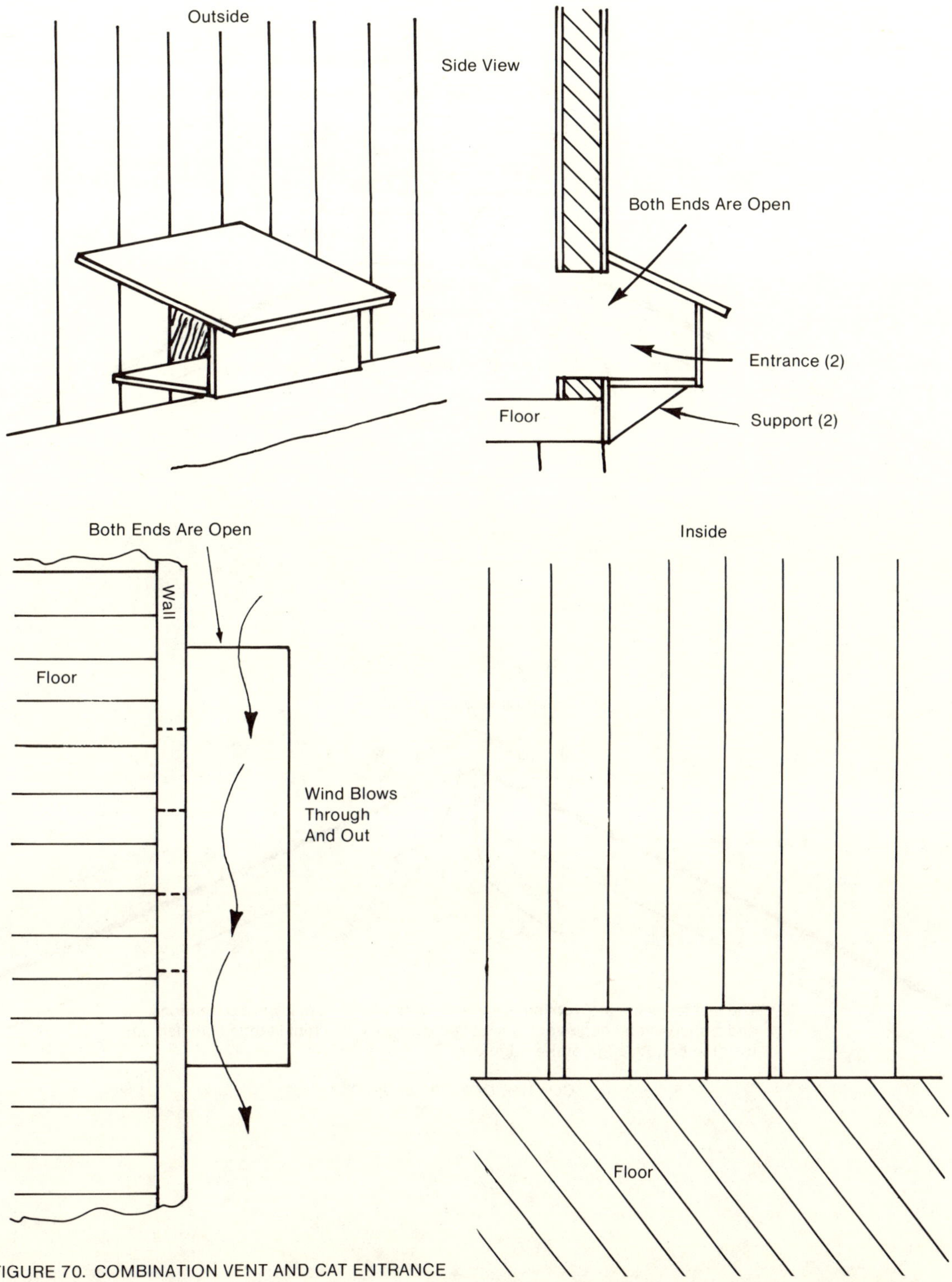

FIGURE 70. COMBINATION VENT AND CAT ENTRANCE

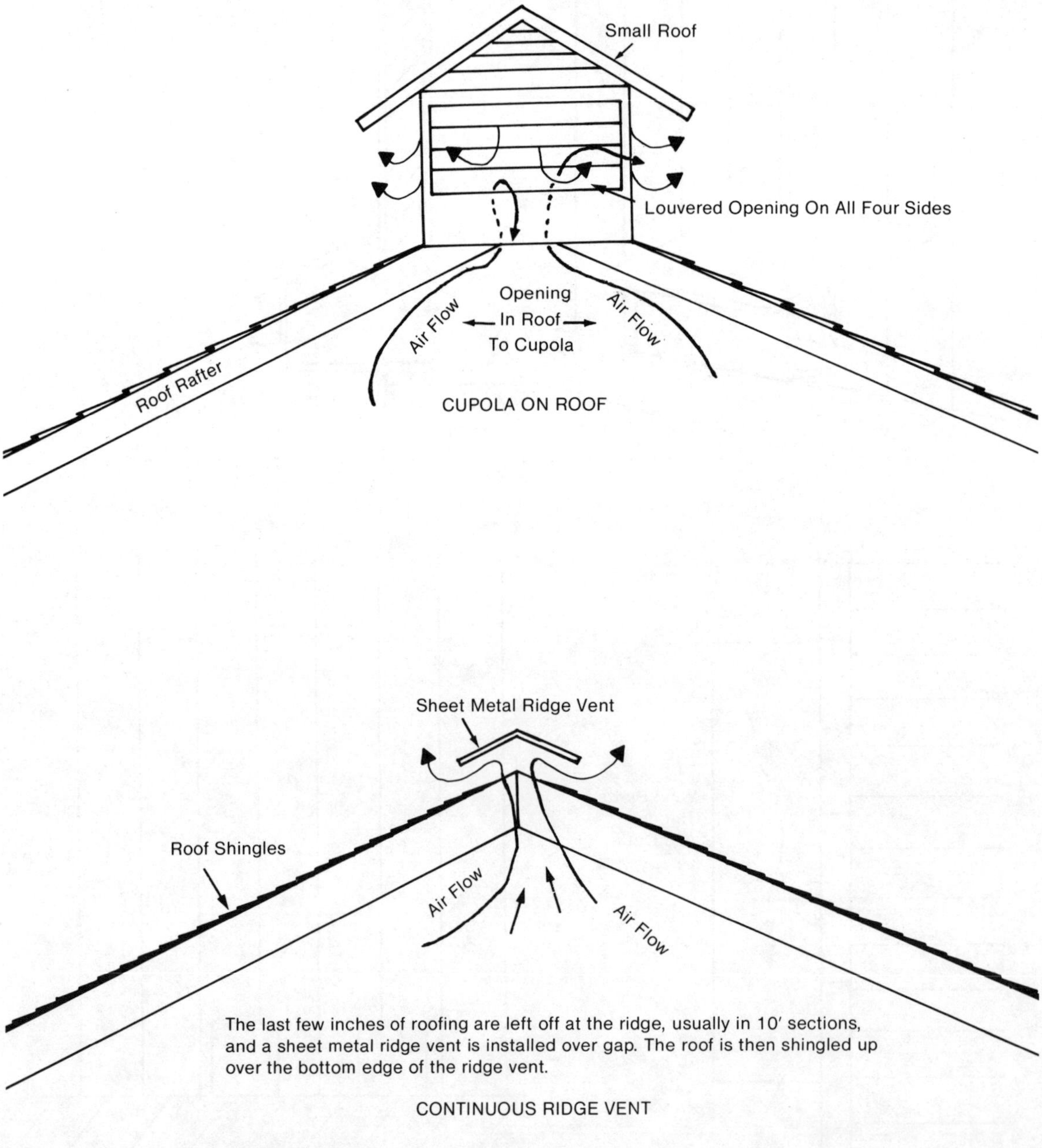
Small Roof
Louvered Opening On All Four Sides
Opening
In Roof
To Cupola
Air Flow
Air Flow
Roof Rafter
CUPOLA ON ROOF
Sheet Metal Ridge Vent
Roof Shingles
Air Flow
Air Flow
The last few inches of roofing are left off at the ridge, usually in 10′ sections, and a sheet metal ridge vent is installed over gap. The roof is then shingled up over the bottom edge of the ridge vent.
CONTINUOUS RIDGE VENT

FIGURE 71.

controllable blast of air which robs an animal of its body heat.

Drafts, although they create some air exchange, do it ineffectively and to the detriment of the animals.

Ventilation creates a controlled air exchange, changing stale air for fresh. Warm, stale air encourages bacterial growth, dampness, and lung problems.

If your animals seemed chilled, check for obscure drafts. When drafts are closed up and your animals still seem chilled, open a window. Damp, stale air makes livestock feel chilled; not fresh, clean air.

More barns need more ventilation, not less. See accompanying sketches for possible ventilation systems in your barn (Figures 69, 70, and 71).

Electricity

Electricity in the barn is a great convenience especially for those times when an animal might have trouble delivering and you need to check her frequently. Chores on a late winter afternoon, a place to plug your electric fencer, the night the vet couldn't make it until after dark—all of these are made easier if the barn has electricity.

Keep the wiring in the barn simple. Unlike the house, the wires are potentially exposed to unplanned interruptions, such as animals reaching a wire, rain getting into connections by running down a wire from the outdoors, etc. These should be guarded against. Still further, your wiring should be exposed so you can check it regularly.

Cover light bulbs with wire protectors. Keep a trouble light and a grounded extension cord in excellent condition (don't use an extension in the barn that's "good enough" or taped where it's worn). When an animal is down where the light isn't the best, use the trouble light.

Heat lamps cause more fires in barns than any other single thing. There are a number of ways to create direct heat that are as easy, and far less dangerous. Heat lamps are often used over a newborn goat kid. We don't use a heat lamp for that. Instead, we dry the kid as much as possible and put the kid in a cardboard box with a bath mat over it. The kid dries a little more slowly but a lot more safely. Piglets also could have a small box to crawl into.

A heat lamp is useful and there are instances where it's invaluable. But, if possible, do the job without one. And when you do need one, be constantly conscious it's on and check it frequently.

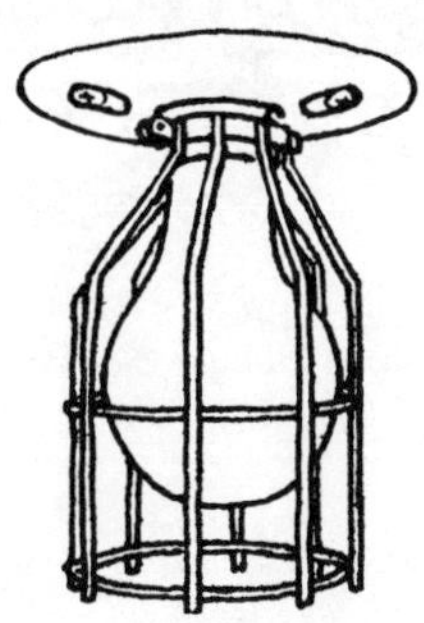

FIGURE 72. BULB PROTECTOR

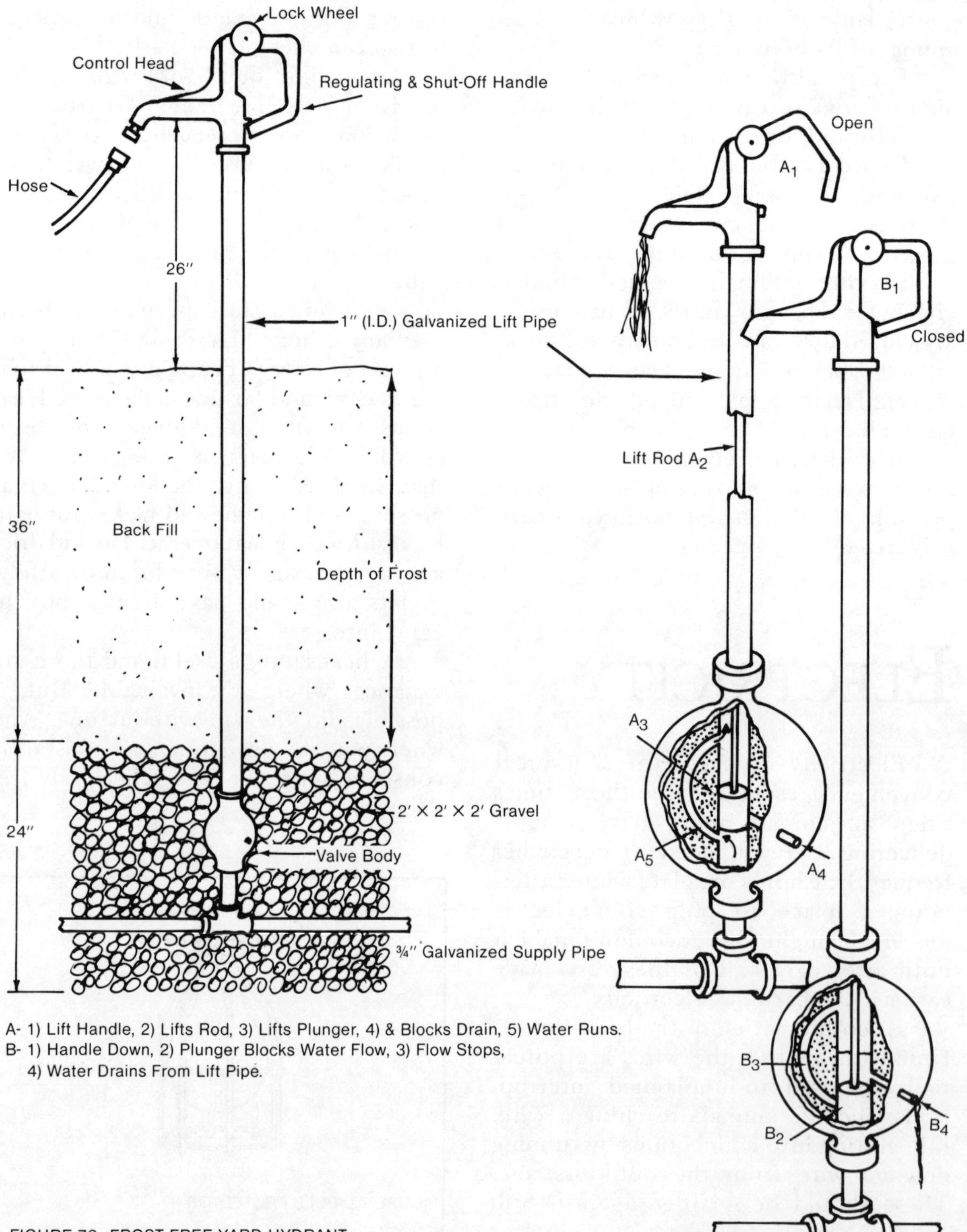

A- 1) Lift Handle, 2) Lifts Rod, 3) Lifts Plunger, 4) & Blocks Drain, 5) Water Runs.
B- 1) Handle Down, 2) Plunger Blocks Water Flow, 3) Flow Stops,
4) Water Drains From Lift Pipe.

FIGURE 73. FROST-FREE YARD HYDRANT

Water Hydrant

What could be more convenient than water in the barn? I've been on the bucket brigade when I had five horses, ten goats, six rabbits, and two pigs. By the end of the day, I marvelled at how much water they consumed. And I marvelled at how many muscles I could develop and still have them hurt. I dreamt constantly of an in-barn water supply and had vague thoughts of $3000 water systems and knew it would never come to pass.

Then I came across a true marvel in a Sears catalogue called a Frost-Proof Yard Hydrant. It taps right off your house supply with a minimum of fuss and, blessings of blessings, it won't ever freeze. If you've had a hose system where in cold weather you spent the day defrosting the hose, that sounds like a miracle. To make it even better, it's low-cost, and simple to use and install.

Firewood Storage in the Barn

Many people store firewood in the barn these days. It seems like a good idea: the roof is already there, and you can pick up a day's supply when doing chores. A few even store firewood in their basements with the same reasoning—and it's even easier to reach in foul weather.

But is it really dry in either place? In the barn, moisture from animals slows the drying process considerably. And many homes have damp basements. Air circulation in both locations is also minimal.

In the "olden days" people built woodsheds. Occasionally, in very cold climates, woodsheds were attached to the house. But the "better" establishments built them away from other buildings and took good care of the structure as well.

Why bother? Firewood needs excellent air circulation, so a "too-well ventilated" building is best. That eliminates the house and *should* eliminate the barn. Dampness—or rather, lack of it—is also important in drying firewood.

But it goes still further. Trees are chosen for firewood because they are crooked or stunted or crowded. They may also be suffering from disease: fungus, carpenter ants, termites, and a wealth of other problems.

While it's generally true these insects and fungi only attack unsound wood, your barn and/or house may have weak spots where these problems can invade the structure. Cutting, splitting, and storing firewood away from your permanent buildings mean these diseases are killed by the cold and ventilation before they are ever brought into the house. Keep no more than a two-week supply of wood in the house.

Stack wood in the woodshed on old tires so the bottom row of splits will be off the ground. Stack the rows so that your cats can get between them for rodent control as well as for improved air circulation. To preserve the building, treat it with drainage oil.

Drainage Oil as Wood Preservative

Drainage oil is still free for the hauling, from both garages and your own vehicles when you change the oil. This "waste product" is very useful in wood preservation. It soaks in well, does not smell badly and, after a month or two, it usually won't rub off if you bump into it. If the building is located in a sunny or windy location, it soaks in even faster.

Waste oil brings out the swirls and burls in the wood, turns it a dark chocolate, and can even be applied over old and flaking paint. For pole structures soak the poles and douse the open ends thoroughly. They will not last as long as creosoted poles but the oil makes a substantial difference. Lichen, water, and bugs cannot get through the oil coating.

If your building is in a very sunny or windy location, put on a second coat (or third) as needed—usually no more than every two or three years. Apply oil with a paint brush or rag. Remember, it's free—so apply lavishly. Spillage cuts the growth of grass at the base of the structure considerably. It means you can't plant petunias there—but it also means no close mowing.

Storing Lumber

While firewood should be stored elsewhere, lumber is often stacked in the barn. Since lumber is sound wood it can be saved safely in the barn for later use.

Several satisfactory methods of storing lumber which don't interfere with the barn's primary function appear in Figures 74 and 75.

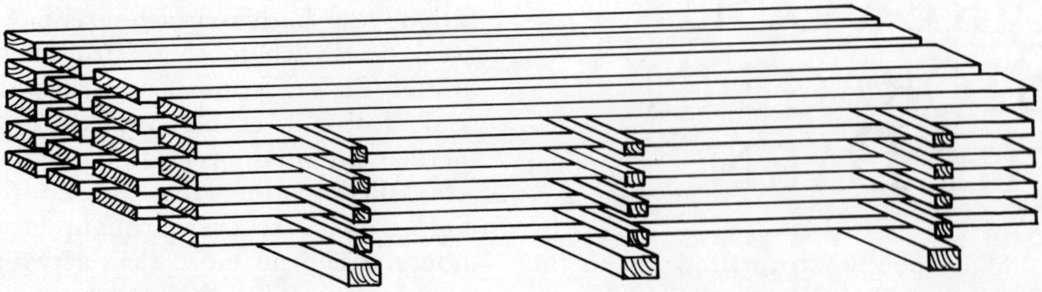

FIGURE 74. DRYING LUMBER IN STACKS

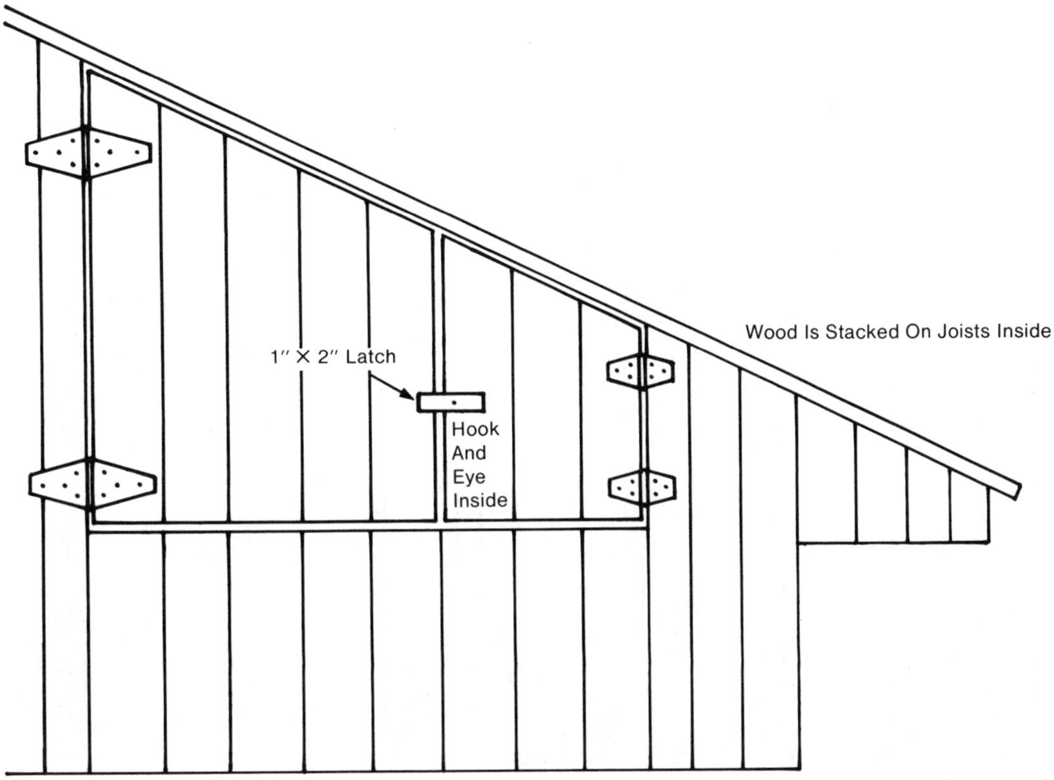

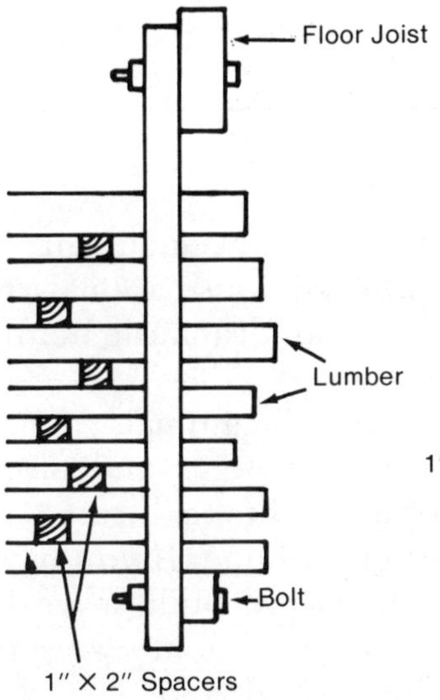

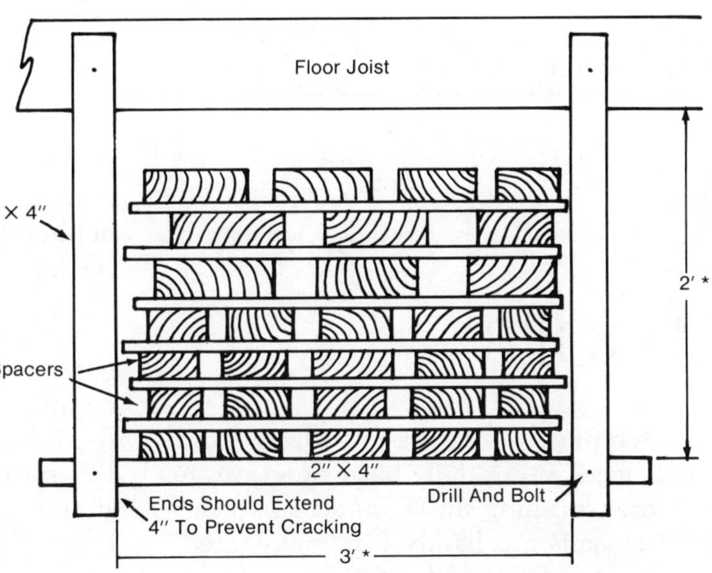

*Any Stack Of Lumber Larger Than 2′ × 3′
May Exceed Strength Of Rack Or Joist

FIGURE 75. LOFT AND BASEMENT STORAGE OF LUMBER

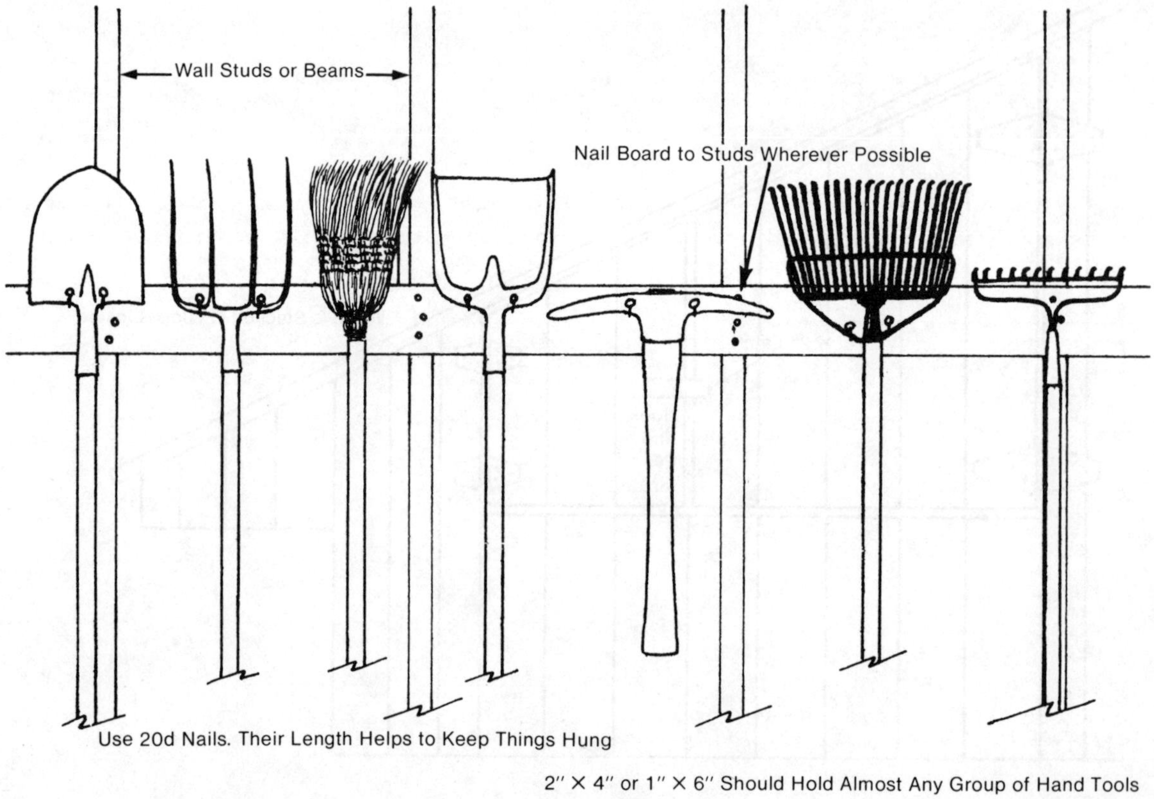

FIGURE 76. TOOL STORAGE ON WALL IN BARN

TOOLS

Scrapers, pitchforks, shovels, and brooms all aid in daily barn function. A hammer, pruning shears, and a small can of 12d nails are handy for unexpected, quick repairs. Store these tools in such a way that animals do not come into contact with them by accident or in deliberate attempts by loose animals to see how much they can dismantle before being discovered.

Arrange tool areas so animals cannot get into them at all. Larger and more complicated tools might be located in the loft or another building. If you have the room and choose to build a workshop in the barn proper, avoid *heating* it. It can be a fire hazard and a health menace to the animals housed there (by keeping them too warm).

Renovating the Older Barn

Don't be hasty in tearing down your older structures. Lumber and money are synonymous when it comes to new barns. You need lots of each. How about time? Are you going to be able to work on it full-time? If not, how many weekends do you *really* want to spend on it? And if you've never done any serious building before, figure on the longest possible time it could take you. Then multiply by ten. (After years of building experience, we still do this, so don't feel your skills are being slighted.)

If you're housing stock in the old barn right now, don't consider taking it down until the new one is finished or you may be selling your stock sooner than planned. (Winter always comes early on the year you build your barn.)

If the old barn is structurally sound, consider adding to it. Or take part of the old barn down and rebuild it to make a better, sounder building.

If you aren't certain the old barn is sound, get two professional opinions. Pay for the advice if you plan on doing the work yourself. Carpenters know their business better than you do, so their advice is worth paying for.

If you decide you must take an old barn down, do so in the reverse of how it was put up. Many people begin without a thought by taking the wall supports down first. If you weaken supports before the roof is down, you put yourself in a lot of danger. And it's much harder to clean up the mess if it collapses. Keep piles of refuse as confined as possible and plan to spend lots of time with a magnet cleaning up loose nails. Animals get "hardware disease" from accidentally consuming bits of metal, not to mention the potential of hoof punctures and tetanus.

Take the time to study the construction methods used; it will help you when building your own structures.

IN CLOSING

An enduring agriculture must never cease to consider and respect and preserve wildness. The farm can exist only within the wilderness of mystery and natural force. And if the farm is to last and remain in health, the wilderness must survive within the farm. This is what agricultural fertility is: the survival of the natural process in the human order.

WENDELL BERRY
THE UNSETTLING OF AMERICA

There's an old country saying, "A barn is blessed with a swallow's nest." As a general rule, it's better to keep wild animals away from your domestic stock. The most important exception is the swallow. They keep flying pests down and raise young that do the same. I always provide them with ways to get in and out, because they must be able to feed their young often. The swallows remind me that wildness is a part of every farm—or should be—and that animals are more than meat-milk-egg-work-producing; they are living beings with lives of their own.

•

The small stock raiser today has lots of problems and few social rewards, but we are growing in number, and society has been noticing. Perhaps the day will come when the social rewards for our efforts at the kindly use of life and land will be given due credit.

In the meantime, we can make do with the satisfaction of doing. Give yourself room to grow and learn from your mistakes. And the very best luck to you.

BIBLIOGRAPHY

BELANGER, JERRY. **Raising Small Livestock.** Emmaus, Pa.: Rodale Press, Inc., 1975

BERRY, WENDELL. **The Unsettling of America.** San Francisco: Sierra Club Publications, 1978.

CARTER, WILLIAM H. **Horses, Saddles and Bridles.** Baltimore: Lord Baltimore Press, 1902.

GALLICO, PAUL, and MATH MATA ANSTALT. **Honorable Cat.** New York: Crown Publishers, Inc.; Don Mills, Ontario: General Publishing Co. Ltd., 1977.

HOLMES, JOHN. **The Farmer's Dog.** 1960. Rev. ed. London: Popular Dogs Publishing Co., 1975.

LEACH, CORL A. **Aids to Goatkeeping.** 8th ed. Revised and compiled by Kent Leach. Dairy Goat Journal. Columbia, Mo.: Tiger Press, 1975.

LORENZ, KONRAD. **King Solomon's Ring.** Translated by Marjorie Kerr Wilson. New York: T.Y. Crowell Co. Inc., 1952.

MACDONALD, JOHN D. **The House Guests.** New York: Fawcett Publishing Co., 1973.

MCGINNIS, TERRI, D.V.M. **The Well Dog Book.** New York: Random House, Inc., 1974.

_____. **The Well Cat Book.** New York: Random House, Inc.; Toronto: Random House of Canada, Ltd., 1975.

Merck Veterinary Manual. 4th ed. Rahway, NJ: Merck & Co., 1975.

MERILEES, DOUG, and EVELYN LOVEDAY. **Low-Cost Pole Building Construction.** Charlotte, Vt.: Garden Way Publishing Co., 1975.

SIMMONS, PAULA. **Raising Sheep the Modern Way.** Charlotte, Vt.: Garden Way Publishing Co., 1976.

SPAULDING, C.E., D.V.M. **A Veterinary Guide for Animal Owners.** Emmaus, Pa.: Rodale Press, Inc., 1976.

STONER, CAROL HUPPING. **Producing Your Own Power.** Emmaus, Pa.: Rodale Press, Inc., 1974.

WHITE, E.B. **The Second Tree From the Corner.** New York: Harper & Brothers, 1935.

______. **The Points of My Compass.** New York: Harper & Brothers, 1954.

Index